INNOVATION IN INDUSTRIAL RESEARCH

I dedicate
this book to
Professor Vijayendra Kumar Garg

INNOVATION IN INDUSTRIAL RESEARCH

PAULO DE SOUZA

PUBLISHING

National Library of Australia Cataloguing-in-Publication entry

Souza, Paulo de.

Innovation in industrial research/Paulo de Souza.

9780643096431 (pbk.)

Includes index.
Bibliography.

Research, Industrial.

607.2

Published by

CSIRO PUBLISHING
36 Gardiner Road, Clayton VIC 3168
Private Bag 10, Clayton South VIC 3169
Australia

Telephone: [+613] 9545 8555
Local call: 1300 788 000 (Australia only)
Fax: +61 3 9662 7555
Email: csiropublishing@csiro.au
Web site: www.publishing.csiro.au

Front cover image by iStockphoto

Set in 10.5/13 Adobe Minion and Optima

Edited by Anne Findlay
Cover and text design by James Kelly
Typeset by Desktop Concepts Pty Ltd, Melbourne
Printed by Ingram Lightning Source

CSIRO PUBLISHING publishes and distributes scientific, technical and health science books and journals from Australia to a worldwide audience and conducts these activities autonomously from the research activities of the Commonwealth Scientific and Industrial Research Organisation (CSIRO).

Mar26_RP_ILS

FOREWORD

Beginning with Ramon y Cajal in 1897, there is a long tradition of books that seek to describe the processes and practice of science. Cajal's *Advice to a Young Investigator*, written years before he won the 1906 Nobel Prize for Medicine, is a classic that is still read appreciatively today. The same can be said of Peter Medawar's 1979 *Advice to a Young Scientist* that appeared 19 years after he shared the Nobel with Australia's Mac Burnet. There are many other worthwhile treatises of this type that discuss the principles and practice that govern a career in academic science.

Conspicuously lacking, though, are comprehensive accounts of what it takes to build a substantial profile in the area of industrial research. That is what Paulo de Souza's *Innovation in Industrial Research* aims to achieve. Of course, the fundamental realities are the same for the academic and industry-oriented scientist: identify significant problems, formulate innovative hypotheses, follow the data, look for the unexpected, tell the truth and so forth, but many of the more practical aspects are different.

Hypothesis-driven academics have, in the past, been driven solely by ambitions that involve publication in top journals, securing competitive grant funding and building a major reputation. Even that, though, has changed in this era of high technology, where a novel finding in, say, nanoscience or from a genomic screen can translate very quickly into practical applications and the possibility of real wealth for the investigator. That can change the way that we both approach our research and report its outcomes. As in everything, there are some simple rules, though without prior knowledge these may often be recognised only in retrospect, and with regret.

Having a book like *Innovation in Industrial Research* to hand can play a part in educating even the most basic scientist to the underlying realities of that interface between the sometime alien worlds of business and scientific research. We haven't worked this relationship particularly well in Australia, though there have been some outstandingly successful examples of how research started in university or CSIRO laboratories has led to major practical outcomes. Even so, there is a very real place for a thoughtful and comprehensive account of what it takes to do industrial research, particularly in the Australian context.

If there ever was a time where scientists were regarded as being part of some high priesthood of knowledge and privilege that requires special treatment, that era is now over. Those who fund research, whether from the public or the private sector, are increasingly focused on seeing real, science-based, practical achievements. As we seek to tackle major problems like the need to develop viable renewable energy solutions, the necessity of maintaining economic productivity in a world that will increasingly mandate sustainable development and so forth, the practice of science becomes ever more complex. The approaches that Paulo de Souza describes can help us to find a way through to improved outcomes, from both the social and the financial aspect.

Peter C. Doherty,
University of Melbourne, 17 September 2009

CONTENTS

PREFACE

I decided to write this book because I believe there is a chance to improve the way industrial research is performed. The basis for innovation in industrial research is in science. Science is more than a body of knowledge, it is a way of thinking, a way we sceptically interrogate how things are organised. Surprisingly, the scientific method is still a strange concept to many young scientists and those working as researchers in industry. This has come about, I believe, because there is a gap in the formal education of scientists and a growing pressure for scientists to produce results. Therefore, I decided to bring out a review of the scientific method using the quality tools familiar to those working in industry to show how it can be implemented. I also thought this text was a good opportunity to discuss ethical behaviour of researchers, and the values required of a good scientist.

I would like to express my appreciation to some of my colleagues for their criticism, comments and encouragement. I received helpful suggestions from Jeremy Breen, Andrew Davie, Meredith Hepburn, Joel MacKeen, Peter Marendy, Andrew Pratt, Peter Taylor and Tim Warren. I would like also to thank Professor Peter Doherty for his generosity and kind attention and the interesting discussions we have had about science.

I am fortunate to have the support of my wife, Anelisa, and the understanding of my daughters, Júlia and Laila, for the many hours I was away reading the literature and writing this book.

I am indebted to my editors Briana Melideo, John Manger, Tracey Millen and Anne Findlay for their outstanding support.

Paulo A. de Souza Jr.
Hobart, September 2009

Part 1

Research, development and innovation

Chapter 1

Introduction

The scientific method is something any researcher can master with daily practice. If it is used properly it can bring fantastic results and guide the process of building knowledge and solving practical problems. It has well-established procedures, while at the same time revealing the researcher's style of doing work. In addition to the scientific method, most of the innovative processes in industry are guided by the application of some quality classical tools. These include brainstorming, Ishikawa diagrams, Pareto graphs, PDCA cycles and mind maps (see Chapter 3). The use of these techniques is common to almost all industrial sectors and hierarchical levels. However, even using the best tools available, many industrial research centres do not identify opportunities in the market; establish priorities in research, development and innovation; secure their investments with patents; or implement non-disclosure policies and practices. Furthermore, some researchers fail to adhere to ethical conduct while doing their work. This text is about all these and other roadblocks affecting the innovation process in industry. I have provided examples to illustrate this. The book is written to help postgraduate students and professionals understand important aspects of their work in research, development and innovation in industry. However, before discussing details of the scientific method, scientists need to discuss why they do research, and to identify the interests of industry in doing research.

WHY DO SCIENTISTS DO RESEARCH?

Many problems can be solved without having to do research. These problems are the ones that have immediate solutions. For example, when it rains (problem) you should look for shelter (solution). Even if the problem becomes a little more complicated, the solution will be immediate. For example, if there is lightning you should stay away from trees and high terrain. Note that you have made use of scientific knowledge: the incidence of lightning strikes is much higher near trees and in high terrain. What if there is a risk of a flash flood, or a landslide? Well, you should still look for an appropriate place in which to shelter! But I don't want to exhaust you so early in the book. By now you will have noted that, as the problem becomes more complicated, you will need to have more knowledge about it; need to collect more data; need to analyse the possible solutions; need to exclude some of those and evaluate to check if your solution really does solve the problem.

Scientists carry out research to solve problems that have no immediate solution. Many researchers[1] put a lot of effort into finding solutions that will increase the quality of products or decrease the production costs in their industry. Other researchers are interested in improving the environmental performance of the organisation they are working for. Improving quality of life is another example that interests many others. In all these cases, the subject chosen for study is called *the problem* and the desired result is *the solution.*

You can use common sense to solve many simple problems. However, common sense is limited; it is based on immediate perception, collective beliefs, experience, personal conviction or even on emotion. Common sense does not submit itself to systematic criticism and is based on opinions. Scientific research rejects the common sense approach.

Research is needed when the problem has no immediate solution. Research is performed using specific criteria – these criteria constitute a classical method, widely accepted as the best available procedure to solve problems. This procedure is called the 'scientific method'. In this book I will explore this method in detail.

There are of course other ways to obtain knowledge. In addition to the scientific method you can use philosophy, metaphysics, psychology,[2] and theology. However, here I will discuss how to solve problems with the scientific method, not with prayer! Science is about explaining the nature of the problem, not blindly trusting that you know what the problem is.

When you have selected your method, think of it as your way of achieving something and look on techniques or tools as the resources you can use to implement that method. (In my view, it is worth describing not only the scientific method but also the tools you can use to implement it.) Techniques I will present in this book are widely used and well known in industry. They are quality tools such as brainstorming, Pareto and Ishikawa diagrams, GUT (gravity, urgency and tendency) analysis, among others.

The scientific method can be understood as a way to generate knowledge from information published on a selected problem; the proper way of obtaining and handling data; the systematic analysis of data; and the use of the information obtained to test a given hypothesis and propose a new thesis that would explain the studied problem.

The value of research in industry comes from a good definition of the problem, a reasonable explanation of its causes and consequences, improvement of processes, cost reduction and so on. There are some companies that carry out research to identify and create market opportunities.

Research requires motivation, innovation, commitment and curiosity. A researcher is a professional prepared to think differently and to analyse alternatives. He or she is someone who has the potential to generate new ideas and do innovative work. Usually researchers have more questions than they have answers. An experienced researcher knows what method to use to obtain answers: very simply – it is the scientific method.

INTERESTS OF INDUSTRY AND PROBLEMS IN RESEARCH

A responsible company wants its business to be sustainable.[3] This means the company wants profit, growth, the wellbeing of its staff, having a good relationship with stakeholders, satisfied clients, and a safe production process with minimum impact to the environment.

All or any one of these motives will drive an industry into doing research. However, even if you have an idea as to what is in the best interests of the industry, it is not an easy task to identify problems to study.

Research performed in industry is quite different from that carried out at a university. Two differences that are immediately identifiable are the time available to obtain the first results and the need for the application of the results. 'Good research' in industry is research that brings results immediately; either explaining anomalies in a process or solving a production problem. There is no doubt the short time allowed for performing research limits the quality of the results obtained. Because of this pressure, less experienced researchers try to 'earn time' by skipping some phases of the scientific method to produce results as soon as possible. This bad practice leads to poor definition of problems, a less thorough study of the literature, and hasty data sampling. This will also lead to poor data analysis, wrong conclusions and poor documentation of the research. The primary consequence of this practice is that the industry's problems will persist.

Poor definition of a research problem is a dangerous habit. Researchers who skip some of the steps of the scientific method are likely to 'go around in circles', and after some time they'll come to believe that all problems are the same. It is important to bear in mind that when carrying out industrial research the time available should be carefully allocated and balanced. Restricting the time taken to achieve results can be a good practice to discipline researchers, but can also lead to bad results. Good managers of research projects should be able to balance the time available with the required quality of the research outcomes.

The industry undertaking research might prefer a good result that can be achieved quickly to an excellent result that takes ages to fulfil. In this case the industry expects its researchers to understand the problems they are facing and to be able to prioritise their work. My advice, if you plan to work in industrial research, is that you must learn the needs of the company as soon as possible. Most industries want good results in an agreed timeframe. An experienced researcher working in industry has to be able to estimate the impact of a proposed piece of research and set up success criteria for this. I can guarantee that this isn't easy. Many eminent researchers at universities or in government agencies could not bear to work under the pressure and constraints required by industrial research for extended periods of time.

In addition to good professionals, industrial research requires some degree of cooperation with universities and public organisations, good project and intellectual property managers, and a working environment well suited for the research activity.

In the next chapters I will discuss what tools are available for researchers, including the scientific method and its limitations, some ways of implementing the scientific method (such as quality tools), the quality of how you measure your data and the management of research projects. Following this, I will discuss secrecy and openness in industrial research, confidentiality and other precautions, scientific documentation and how you can communicate science to the wider society. I will also discuss the researcher's responsibility, ethical aspects of science and values in research. Finally, I will argue the need for innovation in industrial research and 10 possible ways to promote innovation in the way industrial research is being proposed.

Part 2

Tools for research

Chapter 2

The scientific method and its limitations

In this chapter I will be discussing the scientific method. It will teach you how you can apply your own style to arrive at the solution of a problem in academic or industrial research while at the same time practising the scientific method. I will give some examples to illustrate how to go about understanding a problem and how to avoid the temptation of choosing too wide a problem to study. I will also discuss the importance of a good bibliographic survey, and how a literature review supports a better definition of research problems and helps you to propose experiments and hypotheses. Finally, you will gain some understanding of the steps involved in the confirmation of a hypothesis, the establishment of a thesis and the publication of results.

RULES OF THUMB

Imagine if someone offered you the best method ever devised to solve problems. Wouldn't you want to know all about it?

Well, here it is ...

There is one thing common to all good research, no matter what the subject is – its method. The scientific method could be seen as very rigid, but in fact it's not. Take, as an analogy, a chess game. There are rules in this game that limit, constrain and define the way you can move each piece. But then what distinguishes a chess master from a common player like me? Both the Grand Masters and I know the rules. Also, how can different masters have different 'styles'? Understanding the way each player applies their knowledge while following the rules is what makes chess such an interesting game. The rigour of those rules represents freedom for a good player; so it is with the scientific method. In using the scientific method you will be able to apply your own style and strategy and, with practice, you may become a master in research!

I'm going to discuss the rules of the research game. I'm going to show how the scientific method should be applied in industrial research, just like the regulations that govern how the queen and bishops move around a chessboard.

The scientific method consists of well-defined steps. Skipping one of these steps will result in disaster for your research – it's like exposing your king in a chess game. The scientific method classically has the following steps:

- defining the problem to be investigated
- surveying the literature
- formulating one or more hypotheses
- testing your hypotheses
- establishing a thesis
- publishing the results.

Defining the problem to be investigated

Choosing a problem is the exercise of determining what you are going to study. In this first step you should present the relevance of your study subject, define the circumstances in which it occurs, why it affects a process, or why it is the cause of the problem you have decided to investigate.

A good way to pinpoint the subject of your research is to answer the following questions: 'What am I studying?' and 'What motivates me to do this work?'

The literature review

The literature review should not be just a list of papers, reports or books you eventually read and that are somehow related to your work. They should be a critical review of the relevant material you have consulted. The literature review should help you and the colleagues working with you to better define your research problem and to explain the opportunities to be explored in your work.

The literature survey should be the best answer to the question: 'What has already been done about the problem we are investigating?'

Formulating a hypothesis

A good hypothesis should propose one or more solutions to the problem you have decided to study. It can be understood as the best interpretation of the problem, including its primary causes and its solution. A way of defining your hypothesis is to answer the following questions in as much detail as possible: 'What causes my problem and when does it occur?' This is more about understanding the nature of a problem and better defining the problem to be studied than finding a solution.

Testing your hypothesis

Scientists do not accept a hypothesis that cannot be tested. To test your hypothesis you need to obtain data. This means gathering information that would reinforce, modify or deny a given hypothesis. You need to be impartial and understand that your hypothesis will only be really good if it can stand up against all the possible tests you have carried out to reinforce or refute it. If your hypothesis fails, another better one might eventually be formulated. Throughout this processing of challenging your hypothesis, something is improving – your knowledge about a given problem. Believe me, a researcher's knowledge grows significantly through their doubts. You can validate your hypothesis by giving a detailed answer to the question: 'What have I done to test my hypothesis?'

Establishing a thesis

Once your hypothesis has been exhaustively and honestly tested, and eventually modified, you can consider it as your thesis. It is quite possible that your hypothesis has been considerably modified from your original thoughts. In this book, I am working on the premise that you arrived at your thesis when you were satisfied with all the tests you had performed on your hypothesis. Of course, your thesis, once public, will be tested indefinitely by other researchers. More importantly, your thesis will be considered a hypothesis by other researchers.

Publish the results you obtained

Once you have identified and specified your problem, performed a literature review, established and exhaustively tested your hypothesis and then formulated your thesis, you can publicise what you have studied. In doing so you have to describe all the steps you took to establish your thesis.

Science requires demonstration, verification and reproducibility. It does not accept arguments without a reasonable body of evidence to back them up, nor a hypothesis that cannot be tested. In science, all phenomena are assumed to have a defined cause. Phenomena for which no reasonable explanation can exist are miracles, or misinterpreted observation, or misinterpretation of nature.

These are the rules, the roadmap of the scientific method. In this text I will discuss the scientific method in more detail through an example – imagine that you want to understand the economic impact of climate change on wine production in Tasmania.

WHAT TO RESEARCH?

When doing research it is very important to define the problem clearly. Otherwise, you can find yourself having an endless list of problems to solve, or running around in circles not knowing what to do next. When you work in industrial or academic research, it is ultimately important that you should be able to define problems that are really interesting and relevant to study. If your research department cannot produce a good result after a period of time, you may then define a less relevant problem to study, but one that is easier to solve. Selecting a problem for study can be influenced by its scientific relevance (usually the sole criterion in basic research), its potential immediate return for those interested in the solution and how quickly the results can be achieved, or even the likelihood of success. A lot of research is developed at research centres because they foster opportunities for cooperation with other research groups. Others may be interested in opportunities for publishing research outcomes. It is up to you to decide upon the criteria you will use to support your study topic decision. You may consider different reasons: they could range from an order from your chief to client needs. Now I'm going to concentrate on 'how' to solve the problem you decided to investigate. The motivation for your research will be discussed in Chapter 9.

To get back to the example suggested earlier – you have been asked to evaluate the impact of climate change on wine in Tasmania. The problem at hand is to analyse major possible impacts of climate change on the Tasmanian economy, and provide information on how local government could face the negative impacts and take advantage of positive effects on the local economy.

Well, congratulations! You have just taken the first step of the scientific method: you've defined the problem. In this case it is to assess the economic impact of climate change on wine production in Tasmania.

Is your research problem well defined? You may consider that it's crystal-clear, but I beg to differ. A report on the economic impact of climate change in Tasmania would be far longer than this book!

A piece of advice: don't choose wide themes to study. It is professional suicide. I'm going to digress here – for example, defining a subject such as 'tropical diseases' is much too wide. Even if you decide to investigate tropical diseases transmitted by insects, it is still a bit ambitious. Just studying the diseases transmitted by the mosquito *Aedes aegipti* would be theme enough. Wouldn't you agree? Take a couple of these diseases: dengue fever and yellow fever. If you did a short 'brainstorming' on factors involving these diseases it would still have to cover the mosquito's habits, the local population's commitment to interrupt the life cycle of the mosquito in their region, the efficiency of remediation, the poison used to kill mosquitoes and mosquito larvae, the varieties and similarities of the viruses, how much an epidemic event would cost, an analysis of statistical data on obits, the impacts on tourism ... and on and on. What about defining the identification of viruses in mosquito larvae and its implications in different public campaigns for prophylaxes as your research topic?

Choosing this as your research problem you have too much to work with. Just part of this problem can be treated as one or several PhD theses.

Inexperienced professionals in the business of research fall into the trap of defining research problems such as 'tropical diseases' ... and are soon overwhelmed. Something to note here is that public interest has actually defined the problem for you. You can (and should) specify it a little bit further. The next step, the bibliographic review, will help you to better define and understand your problem. Otherwise, you'll get bogged down and probably will never finish your research. Usually what happens is that the problem originally proposed for research is very broad. The example given above is a typical subject for an entire research institute!

To get back to the earlier example – someone may decide to study the impact of climate change on the Australian economy. I'm going to reduce it further and suggest you choose to focus on the impact of climate change on tourism in Tasmania, or even make a detailed study on the impact of climate change on wine production in Tasmania.[1]

WHAT HAS ALREADY BEEN DONE ABOUT IT?

Most of the researchers working in industry forget the importance of a literature review. I am concerned about this because it reveals an important gap in their formal education. Students today are not used to reading widely or undertaking a literature review. I will explain here how a good literature review should be done and will discuss how important it is for understanding problems, as well as for hypothesis formulation. A literature review will also help you to specify your problem better. This is quite interesting work – it's like being a detective looking for evidence of a crime.

To get back to the problem about the impact of climate change on wine production in Tasmania, by doing a 'brainstorming' session you can highlight some interesting aspects of climate change. You may list points such as: inventories disclosed by the industrial sector, uncertainties, certainties, impact, vulnerability, mitigation, forests, sea level, diseases, migratory changes, advantages and risks, to name a few.

At this point you may note that the problem may be wider than you initially imagined. To establish priorities to be investigated in your literature review you can use some quality tools such as GUT (gravity, urgency and tendency) analysis or a Pareto diagram. These and other quality tools will be presented in detail in Chapter 3. These tools are very popular in industry and can be used to help you put the science method into practice.[2]

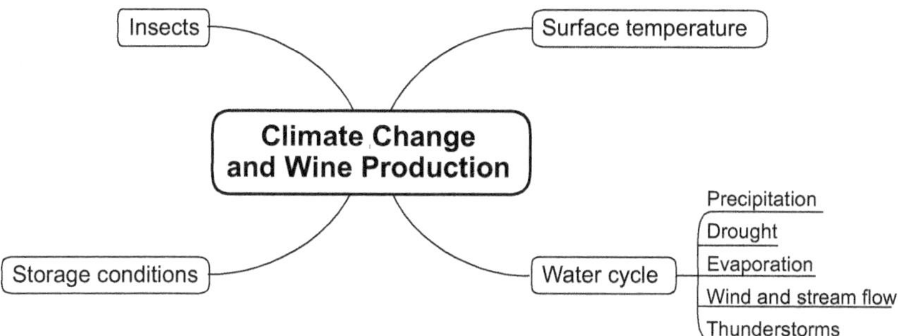

Figure 2.1: Mind map showing variables related to climate change that might influence wine quality.

Perhaps you have prepared a list of keywords, organised in one way or another, for a literature review. If you work in an industry you can look for internal data about it before heading to the closest library or navigating the web using a search engine.

Now is not the time to worry about technical articles. Remember: you have defined the problem, but you don't have enough details to solve it before your retirement!

Someone interested in studying the impact of climate change on wine production may at this point want to specify how many aspects should be considered in the research. With a good literature review a researcher will be able to describe further details on the problem to be studied. To illustrate how vast the impact of climate change on wine production can be, I prepared a small mind map containing some of the variables that might influence the quality of Tasmanian wine. Figure 2.1 shows this mind map.

Once you have defined the research problem, you can go to the closest public library. It is always interesting to do this once you have completed your primary survey. You can visit Google Scholar, Altavista or Web of Science, to mention a few sites.

Now imagine you have a good scientific paper to hand. I'm now going to briefly discuss how to read it. You may want to start by looking at the title, list of authors, affiliation, abstract and introduction, and so on. I don't read scientific papers this way. I prefer checking the title, keywords, abstract and conclusion. If the abstract does not reflect the conclusions, the paper is most likely poorly written. If you think your paper is still worth reading, move on to the introduction, checking the citations on the way. In the citations check where the references were published, when, and if they are recent. Some general indicators of good references, and papers citing them, are:

- journals with a high impact factor[3]
- references with digital object identifier (DOI)[4]
- recent references
- highly cited references (check Web of Science to check that information), and authors.[5]

Points you may consider as indicators of poor references are:

- papers published in very low impact journals
- you have found another paper by the same author(s) with a very similar title
- conference abstract
- incomplete citations (e.g. missing year, pages). Can you trust data published under these circumstances?

At this stage you would be able to judge if it is worth investing your time in reading the paper. If you think it is, go ahead and read it. Make a copy of it so you can mark interesting parts of the text with a highlighter pen. Once you have found what you think may be good reading material, investigate it further. Web of Science and Scopus offer a good way of seeing who cited the paper you're holding.[6] The literature is linked by citations (either citing some authors or being cited by others).

I use a paper notebook, rather than a computer, in which I make notes on papers I read. There, as well as its DOI or reference, I attach photocopies of graphs or illustrations. I also write down any interesting citations for further reading, and conclusions and questions that haven't been addressed by the paper. The important message here is: be organised!

I really like papers that discuss the limitations of their experiments or applied methods. These open questions are true pearls.

Many scientific papers are really bad! Experienced researchers know that rubbishy papers are being published all the time. You must learn how to detect good papers from bad and note which authors to ignore. Other signs of bad papers are the irrelevance of the problem, poor formulation of arguments, poor conclusions, errors in calculations,

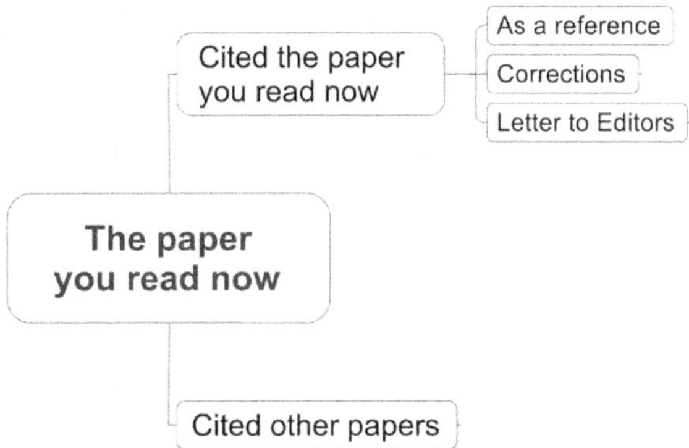

Figure 2.2: Representation of the citation process in a block diagram.

duplication of information in tables, and those containing graphs that do not show proper scale.

Keep copies of papers you read in folders or binders so you can find them easily. This may save you time in the future.

YOUR HYPOTHESIS

Your hypothesis is, in your understanding, the most probable explanation or description of a problem, and frequently presents the solution. The hypothesis is usually formulated based on the identification of a problem, followed by a reasonable literature review supported by your experience and analysis. In this section I will explore common ways to elaborate on a hypothesis. You shouldn't be concerned about validating it at this stage. Your hypothesis should be extensively tested later; if it fails any of your tests, you should refute it, even if it has been devised by an experienced professional. Experience is all very well, but experience without the facts to back it up has no credit in science.

Now imagine you have your problem (or it has already been established for you), and it has been better defined after your literature review because you have become aware of and incorporated new details. Your literature review was useful because it revealed interesting parameters from similar studies. With the better-defined problem and some additional information on hand, you can now develop your hypothesis.

Along with your hypothesis you should come up with ways to defend it against contestation but should allow it to be openly verified or challenged. You will be at a disadvantage because even some simple speculation about something you have not considered or read in your literature review will take credibility from your hypothesis. At this stage, even speculation is enough to put your hypothesis at risk. If you can prove someone's claims against your hypothesis are false, this will not prove your hypothesis is correct. It means that these claims were not able to show your hypothesis is wrong, that's all. You have to understand that, as far as your hypothesis goes, you may have 1000 pieces of evidence to show it is correct, but the existence of just one true piece of evidence against it is all it takes to prove your hypothesis is wrong or not valid in the way you originally proposed.

This sounds cruel? It is – but there is a solution if you consider this one piece of refuting evidence to be an exception. (One case in 1000 ... surely it can't be that rigorous, especially in industrial research!)

An exception or a small deviation may be acceptable in industrial research. Look on exceptions in a positive way, explore them to explain limitations or to improve your hypothesis. To illustrate this, industry has statistical process control.[7] An industrial process is not out of control if a couple of points in 1000 are outside the limits of control.

On the other hand, it is important that researchers do pay attention to detail. Some interesting discoveries have been found in exceptions. Creative analysis on exceptions can reveal new perspectives of a problem. Good researchers pay attention to an exception as an opportunity to innovate and discover the new. The discovery of X-rays and penicillin are good examples of this.

Exceptions are very important in scientific research. They may indicate relevant limitations of a hypothesis. If your hypothesis fails in a particular situation and you cannot explain it, declare your limitation. Be transparent. Other research may provide a reasonable explanation for your limitation, or you may find one yourself in the future. In industrial research, establishing a 'reasonable' hypothesis can be an important result in itself. The 'reasonable' can forecast the productivity and quality of your product with 'confidence'. You may consider this a fuzzy criterion ... however, this probability may be the most interesting expression about the uncertainties on your hypothesis. For example, the mean value of 800 measurements is 250 arbitrary units, with a standard deviation of 25 units. This means that the value you effectively measure with a confidence of 95% is 250, plus or minus approximately 2 standard deviations, i.e. 50.[8] This result can be valuable to a given industrial research area. Therefore, establishing your hypothesis requires you to define your scope, its exceptions, when, why and how frequently deviations may occur.

Looking ahead to our next chapter, note that exceptions can be dealt with using some quality tools. One that I will discuss in detail is the PDCA (i.e. Plan, Do, Check and Act). It is a cyclic quality tool (i.e. for continuous improvement) often used in industry.

Most organisations use quality tools. For example, you can define the relevant problem using GUT (i.e. gravity, urgency and tendency) analysis or, if you already have some measurements, you can use a Pareto diagram. Through them you can address the problem as well as detect other minor problems (or exceptions). You can then specify a new hypothesis and address these minor problems. Just as in the PDCA cycle, you are carrying out scientific research! I will explain these techniques in greater detail later on.

Before moving on I'd like to give you an interesting example. One of the most difficult tasks in biological research is to establish a hypothesis that is 100% valid, and free of exceptions. Take a *description leaflet* on any drug that has been tested exhaustively for many years before being prescribed to you, and you'll read something like: 'the active principle has shown to be efficient to decrease the symptoms in 83% of the patients in medical tests …' Would you take a risk and use this medication? I'm sure you would!

Once you have considered all the data and facts you've gathered on your chosen problem you might be ready to elaborate upon your hypothesis. Eventually you may formulate two or more naturally exclusive hypotheses and decide upon one of them as you collect more information with your experiments. The process of elaborating a hypothesis can be understood as a continuous process. It is a fine-tuning process where your problem can be understood more and more clearly, gaining form and consistency.

In the case of climate change impacting on wine production in Tasmania, after studying the available literature you can state the hypothesis that climate change will be beneficial to some activities (better wines and warmer summers) and will have a negative impact on others (no snow over Mt Wellington).

Establishing a hypothesis is the apex of an intense process of analysing your problem and discovering what has already been done and learnt about it. The better your work in the initial phase of the research, the easier and smoother your job will be towards the end.

Now that you have defined your hypothesis, you should put it to the test!

PROVE YOUR HYPOTHESIS IS CORRECT!

Putting your hypothesis to the test is vitally important and demonstrates the simplicity and the elegance of the scientific method. The intellectual challenge of elaborating a relevant hypothesis, determining its limits and defining the scope of validity is rewarding.

While many consider that the researcher's work can sometimes be tedious, I would argue that each problem offers the opportunity to discover new worlds, a new challenge, and a new chance for learning.

Any hypothesis should only be accepted after it has been exhaustively tested. The validity or refutation of a proposed hypothesis is only obtained via experiments, data, facts and reasoning. If the experiments designed and executed in a laboratory or in the field reinforce your hypothesis, it can be considered, temporarily, correct within the limits of your experiment. For example, imagine you are studying the movement of objects. If they are approaching each other in a straight line with velocity v measured against a fixed observer you may say they are approaching each other with velocity $2v$. Is that correct? Well … don't be disappointed if I say, 'Not always!' If the velocity is close to the speed of light, let's say, 80% of that speed (i.e. almost 240 000 km/s), the objects will approach each other close to the speed of light. At high velocities, close to the speed of light, Newton's mechanics fail. Einstein stated in his hypothesis that the maximum possible speed is the speed of light. Now his hypothesis has become widely accepted – a physical law – or a golden, let's say, first-class hypothesis!

The hypotheses of Sir Isaac Newton are valid within a specific domain that was made clear recently: at small velocities. At high speeds, Einstein's hypotheses should be used.

IS IT A DEAL?

Once you have established a well-defined hypothesis, have performed experiments to verify it and carried out other verification processes you consider relevant, you might be able to test your hypothesis against data and facts you have obtained. If your hypothesis survived this 'baptism of fire', or was backed up by your experiments, you might consider it as your 'thesis'. (At this point I don't want to delve deeply into areas that have been thoroughly examined by philosophers, so I have put interesting texts from Popper, Descartes and others in the bibliography for you to read, should you want to know more. In the context of this book I will treat the thesis as a well-proven and eventually accepted hypothesis.)

In any thesis its universe of validity is as important as the thesis itself. If you are planning to investigate the positive impacts of climate change on Tasmanian wine production, you should define which aspects you are willing to study. For example, you may consider investigating how warmer summers and winters will influence Tasmanian grapes and the quality of its wine. Will it affect the wine quality favourably? If so, is it true for all grapes and wine types? You have to specify what you mean.

Therefore, just as important as determining the validity of a given hypothesis is to know the context in which it is valid. As far as you are concerned, your thesis is established once you have confirmation of your hypothesis. But for other researchers your thesis will be a hypothesis that they are going to test. No matter how much you have tested your hypothesis and how convinced you are about your thesis, there is no guarantee it is always valid ... you should always bear this in mind. It is worth repeating: *your thesis is considered as a hypothesis to be tested by other researchers.*

SPREADING THE NEWS

I'll review what's been discussed.

- You have established a topic to study (your problem).
- You have reviewed the literature about it.
- You have revised your problem.
- You have elaborated a hypothesis and its limits.
- You have tested it and established your thesis.

Now it's time to publish the results!

In the next part of the chapter I will discuss reasons for publishing the activities or results of industrial research, and give many other reasons why you should take care while doing so.

I'll start with some reasons *for* publishing your work.

1 Peer-review process

When you submit your work to the peer-review process, you will receive comments supporting your thesis or expressing key doubts about its validity. Being exposed to these situations is a good enough reason to write a technical paper and submit it to a scientific journal.

Having other (and perhaps more experienced!) researchers evaluating your work is a very important step towards giving you more confidence in your own work. Having a paper accepted for publication is also a small step in building your reputation within your professional community. Once published, your work will be read by interested researchers and they will act as your peer reviewer, reporting their findings in different ways: citing your work (for good or bad reasons), writing to you about their assessment of your paper, writing to the editor about it (usually when they have reasonable arguments against your work), nominating you for a prestigious award – who knows! Many industries do not submit research to external peer review. In this case they should establish an internal peer-review process.

2 Contribute with future research

The same way you consulted and carefully read papers published in the literature, and these articles supported your research, you can also contribute to others' work. This is an important mechanism in science.

3 Prove your experience in the field, when necessary

The best business card a researcher can have is their published papers. This is even better than a line in their curriculum vitae stating years of experience in a research centre. However, it will always be a challenge to judge a researcher's real technical skills without submitting them to an interview or technical discussion. What a researcher or research centre publishes is understood as evidence of what they have developed or achieved. In principle, the more you publish, the more productive you ought to be. I will discuss science health indicators later in this book.

4 Create or contribute to your organisation's technical memory bank

The modern organisation should take care to collect and store technical documents relevant to their business. It is common practice for research groups to make their published material available for visiting researchers or students to consult. You can contribute to the dissemination of knowledge in your organisation if you help to create technical documentation or support an existing department (i.e. library, webpage with list of publications). Some of the technical documents might be classified in accordance with the organisation's restrictions on circulation. For example, if you have worked with a client to solve an important developmental problem on a new product you should avoid making it public, unless it is in the interests of your client and your company to publicise it. Strategic documents should have a very restricted circulation. On the other hand, there are many other interesting documents you can publish without putting the sustainability of the business at risk. I will discuss this in further detail in Chapter 6.

5 Technical marketing

Through its research results, an industry can build or sustain commercial strategies. This is because publications demonstrate to clients that the organisation has the necessary technology and the most up-to-date knowledge about its products, their uses and processes. In this way, publishing the results of work developed with clients or stakeholders could also help your company form a good relationship with the market or community. Each publication, whether with clients or interested sections of your business, should be celebrated!

6 Exercise your ability to compose technical articles

Most technical magazines have very restricted space. This means you only have a few pages in which to describe the relevant aspects of your work. Writing technical papers is therefore an exercise in synthesising complex ideas and obeying very rigid editorial guidelines. Researchers who write many scientific papers usually develop valuable skills in synthesising ideas and in text composition.

I hope I've convinced you of the personal and institutional benefits in publishing your research work! Now I'll give you reasons to *not* publish it …

7 Secrecy is important at this stage in your business

Many industries keep their research activities secret. Employees have explicit instructions to keep quiet about what they are working on. Some organisations request researchers not to discuss their research work or interests in public restaurants, airports, or lifts. Such concern is common in extremely competitive businesses. Other researchers may have some reservations about publishing a result before a patent has been issued.

8 The research needs additional work

After doing your research for some time you may come to realise your work has not achieved reasonable results or that the results cannot be fully explained. This is not a dead end. There are other benefits from research as I will discuss in Chapter 8. However, it is a sufficient reason not to publish a paper.

9 One of your colleagues is against publishing the research

Research is frequently a team effort. If you have participated in research involving other researchers or engineers – whether from your organisation or outside – and one of them does not agree with publishing the paper, you should stop right there. Try to understand why the colleague has reservations about publishing and discuss it with the other co-authors. If their position is firm, even if you do not agree with their arguments, don't publish!

10 It is not a relevant problem

You have to be the best critic of your work. Ask yourself what your research contributes to your field of expertise. It is sometimes difficult to judge how much your paper contributes to the advancement of your research area, but it is not that difficult, honestly, to judge when the contribution is null. Irrelevant work results in publication in small impact journals, a poor citation record, and boring lines in your curriculum vitae.

11 If there is a contract or another legal instrument against publication

Don't even consider the possibility of not adhering to a non-disclosure agreement. You mustn't fool yourself into thinking that a paper published in a journal in another country will not be read by a client or research partner.

I have seen researchers find themselves in a very difficult situation because they have not adhered to non-disclosure agreements. Your reputation is worth more than a good publication.

12 The result is important, not the publishing

Writing technical publications is a time-consuming activity. Some companies argue that they would prefer their scientists did research rather than write papers. Depending on the quality of the work and the audience of the publication, it might be fair to not publish the result.

But what to do then: to publish or not publish research results?

The best answer to this question is the researcher's stock reply: 'It depends …' You must discuss it with your co-authors and supervisors, consult your organisation's code of conduct regarding publications, and then decide. Some of the results you've obtained may not be so important, but the method you developed might be. The way you report how you and your colleagues applied techniques and solved specific problems might be interesting to other departments in the company, as well as to clients and other research centres.

A publication can be considered technical merchandising. As a result of your publication, you could be invited onto interesting projects, to give talks or review someone's project or paper.

Once you have decided to publish your research, find the appropriate journal – consider things like the impact factor, the audience, space available, costs, and so on. Read a few articles from the journal before you start writing the first line and … good luck!

Here is a final and important piece of advice. When you are about to publish a paper or a report, ask yourself – does it claim more than can be supported by your data? In research, scientists seek the truth; this is an extremely important ethical issue. Unfortunately, there are cases of mistakes or fraud in science. I'll mention one serious case here – a few years ago, an editorial from the journal *Science* attracted the attention of the scientific community for a situation involving a renowned young physicist and liar.[9]

The 32-year-old physicist was working in the USA for a globally recognised laboratory, and was the author of many research papers in the prestigious *Science* and *Nature* journals and had been considered as a potential Nobel Laureate. He was questioned about results from his research that could not be repeated by other researchers. The editorial I mentioned was the end of a very difficult time for the renowned laboratory and the scientific magazines. It started a serious discussion about the manipulation of scientific results and erroneous interpretations of scientific research.

Over several months, a committee of five experienced researchers investigated claims that the young physicist had lied in his research reports. Numerous scientists reported difficulties in reproducing the experiments, raising 'serious concerns'. The committee investigated allegations that graphs were falsified, unrealistic precisions were reported and that the results contradicted the known physics. The physicist was accused of the most serious sins in the science world: falsification and data manipulation.

The organisation where the researcher worked (which gave us the transistor, the laser and the operational system UNIX) lived their worse nightmare in eight decades of existence. I can imagine all researchers working there were somehow affected. The important magazines that had accepted his material for publication also experienced very difficult times.

The peer-review process of scientific papers is not perfect. It is assumed by scientific journals that there is a minimum level of professionalism from the researchers who submit their work for evaluation. However, the review process will never guarantee immunity against well-constructed fraud. Some publishers are moving to guarantee minimum ethical standards in their publications – an example is the Elsevier ethics resources kit – and new services are available to capture duplication and plagiarism.[10]

You can contribute as a researcher working to improve the quality standards of your own report, and of those working with you. You have an inexcusable obligation to act if something is claimed that is not supported by the data in the research conducted by a colleague. I will discuss these issues further in Chapter 7.

Unfortunately, cases like this are as common in science as they are in many other sectors. Factors such as the pressure for fantastic results, production or convenience might be pushing scientists towards this unacceptable behaviour. This kind of misbehaviour does not surprise people in other professional areas any more. I would beg you to bear this in mind while you are writing your papers and carefully consider the rest of your career.

In this chapter I have explored formal publication channels. There are many other media available, such as press coverage, your organisation's intranet, educational and outreach activities, to mention just a few. I will discuss them in more detail later in this book. Right now I'm going to set out the most common mistakes you can make that can be damaging to your career (this is even though you might be trying to implement the scientific method):

- skipping one of the steps of the scientific method
- not defining a subject of study well enough
- neglecting the importance of undertaking a good literature review
- making your literature review merely a simple list and short description of published material
- not establishing a hypothesis
- protecting your hypothesis from honest and open criticism
- publishing when you have a non-disclosure agreement; not informing your co-authors about the publications; including authors who have not made a real contribution to the paper
- publishing results you wish you had obtained.

LIMITATIONS OF THE SCIENTIFIC METHOD

The scientific method is a powerful tool for solving problems and spreading knowledge. However, it has its problems. I will list these in five major categories.

1 It is extremely dependent on researchers
   - Peer-review process does not guarantee quality; especially the quality of the data
   - Too much trust in authors. There is no audit process widely established to verify the quality of data, although there are efforts to increase the responsibility of co-authors
   - Duplication, plagiarism, data manipulation, negligence and other misconduct in science is present.

2 Slow process of gathering and disseminating results
3 It is expensive and time-consuming
4 Outreach has been inefficient
   - Inefficient communication strategy
   - Few researchers effectively engage in educational and outreach programs
   - Perception from society shows concern about research limits and practices of science.
5 The benefits do not extend to everyone, especially people in poor countries.

These are important aspects that the next generation of scientists has to deal with and they should be actively considered by today's researchers. These limitations cannot be ignored because they result in poor research standards.

Chapter 3

Quality tools and the scientific method

People only see what they expect to see. I can only imagine how indigenous peoples interpreted the arrival of the Europeans in their ships … most certainly they did not recognise them as large European ships, but something completely different. This is why associations and examples are so important when explaining something new.

The quality tools used by those working in industry are similar to the scientific method in many aspects. These tools were developed in the early 1960s and improved practices in industry across Japan and the USA first, then gained credibility during the 1980s around the world. Today, quality tools are part of normal practice in industry.

Because of their widespread use in industry, I will discuss what quality tools look like when they are used in the scientific method. In my view the scientific method can be strengthened and implemented well using a combination of these tools. In this chapter I will also discuss the conditions that favour innovation in industrial research.

QUALITY TOOLS: VIRTUES AND USES

This section will present some possible uses of well-known quality tools (such as brainstorming, GUT tables, Pareto charts, Ishikawa diagrams and mind maps). While I argue that quality tools are interesting and useful for supporting the implementation of the scientific method, I will also consider their values and potential limitations.

Quality tools are used to assist with the generation of ideas, the treatment and classification of data, defining the most important actions, investigating causes of problems, building a better understanding of phenomena and processes, and supporting decision making.

Quality tools became extremely popular during the 1980s and 1990s and because of their practical value and simplicity are still widely used in industry today. The point here is that researchers can use them to support the implementation of the most powerful method of analysis, identification and solution of problems: the scientific method. Researchers can use quality tools to put the scientific method in place while searching for solutions and a well-structured explanation of problems in industry. In this system, there will be opportunities for innovation.

Brainstorming

What is brainstorming?

Brainstorming is a process of collecting words that have a relationship with the problem under scrutiny. A group of people (it is even better if it is a multidisciplinary group), during a short time interval (e.g. 15 minutes), selects the words coming to their minds about a problem and writes them down on paper. The final list of words can be collated by the secretary of the meeting. No idea can be ignored, criticised or justified. In fact, 'crazy ideas' should be encouraged. Brainstorming prioritises the quantity of ideas (or words) and not their quality. Later the ideas should be organised in sets or lists. A brainstorming session with my colleagues once produced very interesting ideas about the research I was planning to do. However, be careful with the time you allow for it: brainstorming longer than 20 minutes can be considered occupational therapy!

Where to use brainstorming?

Brainstorming provides a list of interesting keywords for a bibliographical survey (but, remember, it should not replace it!), helps in gathering ideas, and provides good material for other quality tools. For example, the ideas collected in a brainstorming session are very important in a GUT (gravity, urgency, and tendency), at the beginning of an Ishikawa

Agriculture; Animals; Bushfires; Dangerous; Drought; Economic impacts; El Niño–Southern Oscillation; Enteric gases; Evaporation; Expensive;	Extinction of endogenous species; Fear; Floods; Humidity; Hurricanes; Inevitable; Land temperature;	New energy matrix opportunities; Precipitation; Rain regime; Rise of sea level; Risks; Sea surface temperature; Seasonal variation;	Snow; Surface temperature; Thunderstorms; Terrible; Urgent action; Water cycle; Wind and stream flow

Figure 3.1: Results of a typical brainstorming session around the impact of climate change. Terms are ordered alphabetically.

diagram, in the preparation of verification forms (see page 36), and so on. So, it will be useful throughout the process of exploring the possible solution of a problem, including the formulation of an alternative hypothesis so that other researchers can follow your work.

Mind map

What is a mind map?

A mind map is a graphical and hierarchical organisation of related items. You can use this tool in organising ideas from a brainstorming session. It is a diagram used to represent ideas around a central topic, word or concept.

Where to use a mind map?

A mind map is very useful for obtaining and organising ideas. It is very effective at capturing creative thinking and presenting ideas.

GUT

What is GUT?

GUT stands for gravity, urgency and tendency. You can use these three criteria to assess the relevance of an idea to the problem you are studying. Sometimes, when ideas are not an issue, the problem is to determine what is relevant and where to invest your resources. It helps you to set up priorities in your work.

To implement a GUT you can rank ideas from 1 to 5 for each criterion (i.e. gravity, urgency and tendency) in a table. Then you multiply the numbers you have allocated under the headings of gravity, urgency and tendency. The higher the result of this product, the higher the relevance of the idea you are considering.

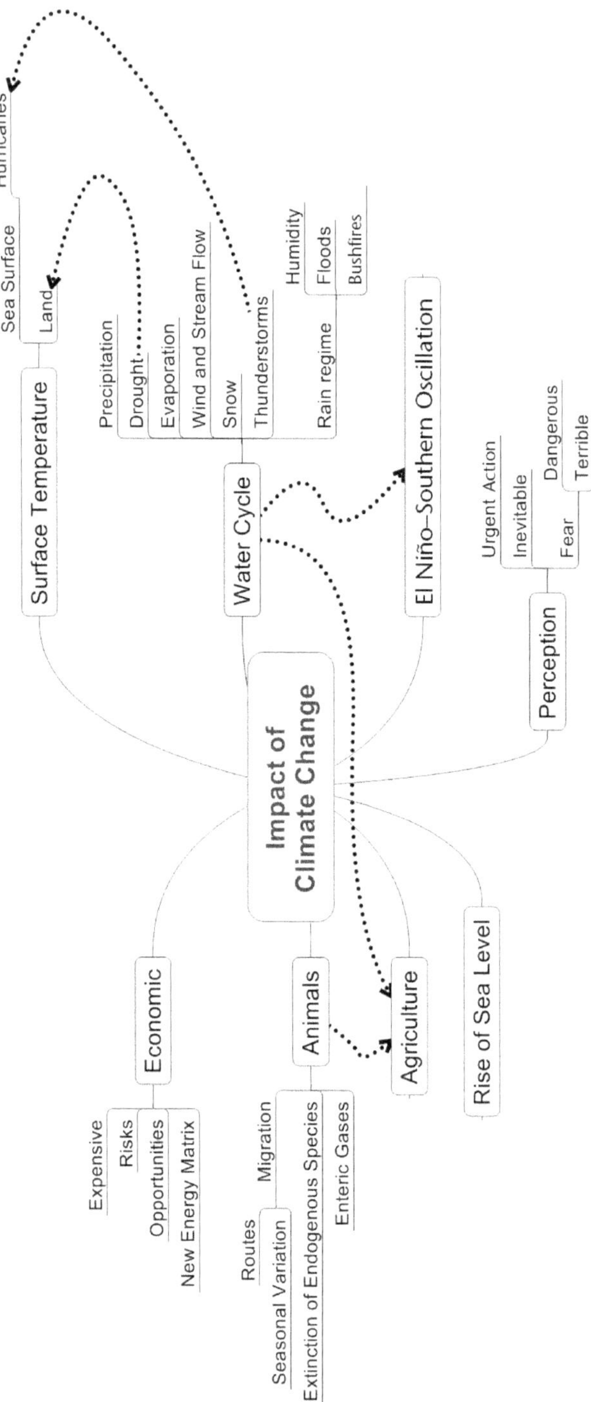

Figure 3.2: Mind map of the ideas collected in the previous brainstorming session (Figure 3.1). In this mind map the ideas were organised into topics.

Table 3.1: A typical table assessing gravity, urgency and tendency of the impacts of climate change in a given region. In order to avoid too much detail here, all aspects are considered to be negative. This means that increases in all these parameters will create a negative impact.

Topic	Variable	Gravity	Urgency	Tendency	Assessment (G × U × T)
Surface temperature	Sea surface	3	2	5	30
	Land surface	4	3	5	60
Water cycle	Precipitation	5	3	5	75
	Drought	5	4	5	100
	Evaporation	4	3	3	36
	Wind and stream flow	3	2	3	18
	Snow	1	2	5	10
	Thunderstorms	3	1	4	12
	Humidity	1	1	1	1
	Floods	4	3	5	60
	Bushfire	5	3	5	75
ENSO	El Niño–Southern Oscillation	4	3	4	48

Where to use a GUT?

You should use a GUT just after a brainstorming session to judge which ideas are most relevant to your problem.

In the example given in Table 3.1 I used some of the variables collected in the brainstorming session (Figure 3.1) and organised in the mind map (Figure 3.2). All impacts should be considered negative and the impacts of each variable should be solved.

The assessment you make may change depending on the relevance of the variable to your process or region. For example, if you are living in a tropical region you may consider an assessment of snow of G = 1, U = 2, and T = 5 (product 10). But if you are in Switzerland or the mountains of New Zealand, being dependent on snow for tourist activities you may consider an assessment of G = 5, U = 5, and T = 5 (product 125).

Cause–effect diagram

What is a cause–effect diagram?

A cause–effect diagram (also known as an Ishikawa diagram) is a graphical representation, developed by Kaoru Ishikawa, that allows you to identify, highlight and enumerate some of the possible causes of a problem. There are different structures you can use for constructing a cause–effect diagram. Here I will present two possible layouts. The first is applicable to problems in industrial plants and the other to equipment itself.

When considering problems in industrial plants, the cause of the problems can be analysed from two sources: management or operational causes. These management causes are known as '5P': policies, procedures, prices, people and plant. The operational causes are known as '5M': method, material, machinery, maintenance and manpower. Once the main causes are identified, you can move on to secondary causes.

These 5P and 5M layouts are two classical initial structures of cause–effect diagrams. However, you can adapt your structure to be more appropriate to your problem. The important aspect in these diagrams is that they search for the primary cause of a problem. If you try to solve a problem without looking at its causes, but only its symptoms, it will be like trying to cure an infection using cold compresses. You may temporarily decrease your fever, but not fix the cause of your infection.

Where to use a cause–effect diagram?

Cause–effect diagrams are designed to help you find the basic causes of problems. For example, relocating people or machines should not be suggested if the problem of a given procedure is clearly related to maintenance. Cause–effect diagrams are known in industry as a quality tool that supports effective problem solving by investing time and resources into discovering where the real causes are, not investigating effects.

Often when drawing up a cause–effect diagram you can assess how much understanding there is about a problem. It may reveal the need for more data (i.e. either within a better level of confidence, precision, or completely new data). Frequently during the construction of these diagrams it is possible to expose deficiencies of data handling in the company. It is considered a powerful quality tool technique and can be applied to a wide variety of problems. Figure 3.3 is the cause–effect diagram constructed in the analysis of a few parameters affecting the quality of wine.

Verification form

What is a verification form?

Verification forms are commonly drawn up as a spreadsheet or a structured form where data can be recorded and easily accessed later

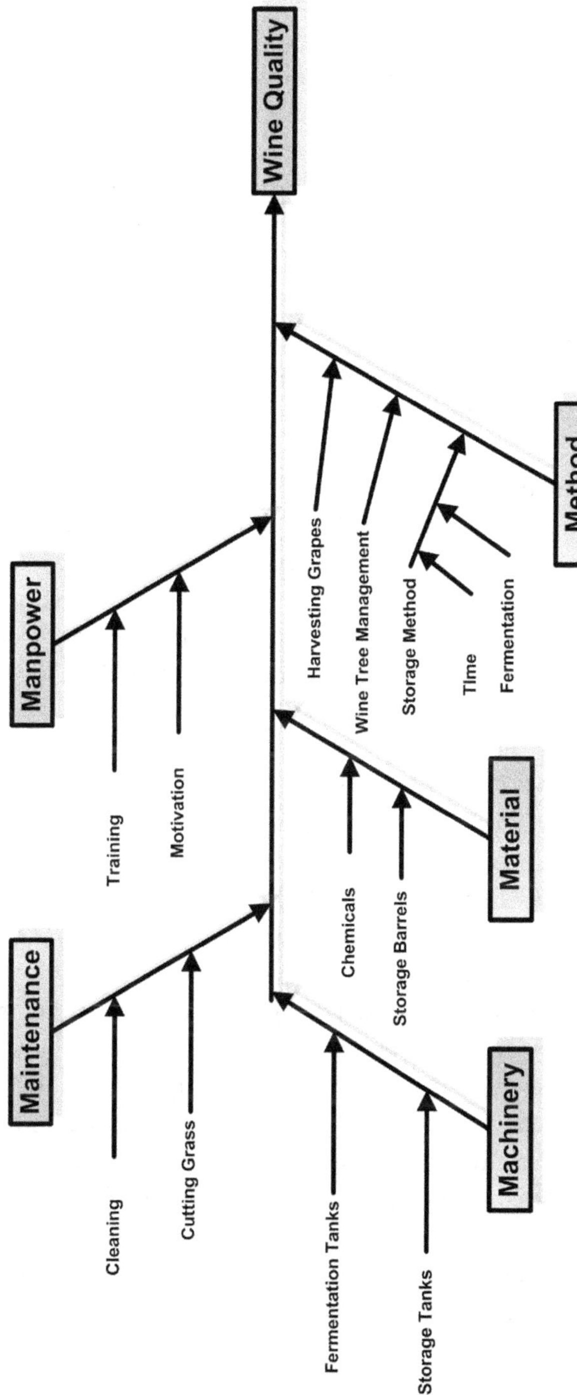

Figure 3.3: Cause–effect diagram of aspects potentially influencing wine production in Tasmania.

for analysis. This data could help researchers avoid problems and could also be used to explain them. Data recorded in verification forms should be selected from a brainstorming session followed by a GUT. When studying a given industrial process there are many statistical tools to help you select the variables you want to have on your verification forms. Once well crafted, verification forms can be very useful in designing and supporting the proper execution of experiments.

A few hours before the launch of a Space Shuttle, astronauts are kept very busy referring to details on a verification form or checklist. Pilots follow the same routine prior to take-off. They are simply using verification forms designed to confirm all the important variables to do with a safe flight.

It may seem an obvious thing to do, but many experiments at universities, research centres and in industry are being carried out without a single verification form! A lot of time and resources can be saved with a simple verification form because by using it you will guarantee that relevant aspects of the experiment are being verified and recorded. This practice prevents you repeating experiments and spending time searching for information on experiments you did a few months ago.

Applications of verification forms

Verification forms are considered quality tools that support the planning and execution of experiments. They can also be used to support the further analysis of problems. They are absolutely necessary while operating complex and expensive machines or executing processes where your safety could be at risk.

Verification forms can be used during data collecting and its immediate analysis, when interpreting problems related to failure (e.g. in production processes or machines), as well as supporting the classification of experimental problems (e.g. mistakes in reading and writing, accuracy while weighing a sample). On top of that you can add documenting processes and experiments for a future investigation. Verification forms in portable devices (e.g. PDAs, mobile phones) are very useful and convenient, while database, data management and different communication technologies are available (e.g. Bluetooth, Wi-Fi, 3G).

Pareto diagram

What is a Pareto diagram?

The Italian engineer and economist Vilfredo Pareto analysed the distribution structure of wealth and discovered that one-fifth of the population holds the majority of a country's wealth. The graphical plotting of this distribution became known as Pareto's diagram. It also became clear that his diagram could be applied in many fields of expertise.

Applications of a Pareto diagram

Basically, Pareto diagrams are used to reveal how important contributing factors in a problem are. Defining relevance of causes helps you to prioritise investments (e.g. your time working on a problem).

To construct a Pareto diagram, you should consider the possible causes of a problem, compare them against a given metric and establish an order of priority. For example, in wine production you may consider that you lose quality when barrels are too young, storage areas are too cold, too hot, too humid or too dry, and so on. You can compare them by ranking the influence on quality or on price.

Your metric can be the frequency of occurrence of the problem, number of claims from clients, increase in cost, etc. They should be considered within a given time interval (days, months or years). At the end you can organise the data in a table listing the causes and results against the metric within the considered period. For example, Table 3.1 considers more than one metric for comparison.

Figure 3.4 illustrates the result of a greenhouse gas inventory performed in 110 industrial installations. The entire emission is of about 9000 tCO_2e/year.[1] The accumulated emissions of 20 industrial installations correspond to 80% of the emissions. If you are planning to reduce greenhouse gas emissions, where should you concentrate your efforts?

Statistical tools

Statistics are used to give meaning to data. From statistical analysis and a correct interpretation you can demonstrate correlations that are not coincidence but a consequence of the interaction of phenomena.

There are many statistical tools that may help you define priorities in industrial research as well as understand the phenomena you are

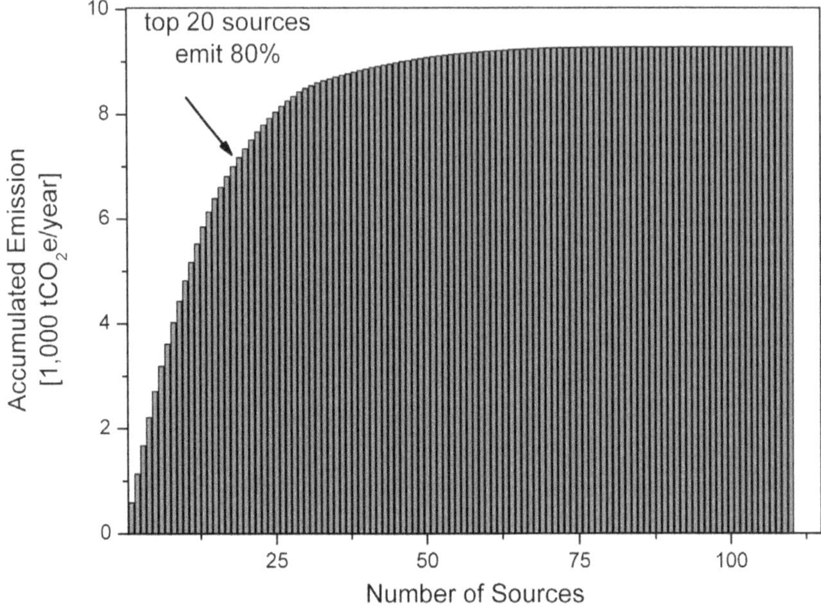

Figure 3.4: A Pareto diagram constructed from greenhouse gas emissions in 110 industrial installations. In this plot we can see that the accumulated emission is about 9000 tCO_2e/year. Just 20 sources together are responsible for 80% of all emissions.

studying. Usually there is an abundance of data in industry, and frequently this data is of poor quality. Nevertheless, you can build dispersion diagrams, control plots, box plots, co-variance analysis and time series modelling. Some of these and other statistical tools will be presented and used later in this text.

Recursive tools

There are many recursive quality tools you can use to implement part of the scientific method. A typical quality tool designed for continuous improvement is the PDCA (i.e. Plan, Do, Check and Act). The idea behind this tool is that a process with a problem can be continuously improved. Here is what you do for each stage of the PDCA cycle in a research-focused activity.

- (P) 'Plan' the new design of your experiments by finding out what needs to be improved, and propose new ideas for solving these problems. Do not forget to estimate what the result of the change would be.

- (D) 'Do' changes designed to improve the experiments (start with easy steps).
- (C) 'Check' whether the expected results were obtained.
- (A) 'Act' to implement changes on a larger scale, reproducing the success of a small-scale experiment. If it is a process, promote changes in the routine and control systems. If it is to do with people's behaviour, you should invest in training.

Of all the quality tools, the PDCA is the one that most resembles the scientific method as a whole, while the other tools are like 'pieces' of the scientific method.

Table 3.2 below organises some of the quality tools in accordance with the support they can give you during the implementation of the scientific method.

It is possible for scientists and researchers to implement the scientific method using quality tools. Because these tools are generally familiar to people working in industry, it would be a straightforward step to take (Table 3.3).

Table 3.2: Overview on some quality tools, their strengths and possible uses.

Quality tools	Strengths	Possible uses
Brainstorming sessions	Uses experience of different people Creative process	Throws up many options to be considered when aiming for a solution to a problem
Flowchart	Empirical Organises information	Supports determining the limits of a hypothesis by establishing the limits of a process
Stratification	Correlation between data Supports statistical analysis	Assembles data sets based on their relevant properties
Histogram and box plot	Real data is used	Determines tendency and dispersion of data
Ishikawa diagram	Creative	Formulation of hypothesis Infers the different causes of a given problem
Pareto diagram	Organises problems under given criteria	Supports prioritisation
GUT	Selects issues to be investigated from a brainstorming session (followed by a mind map) considering the Gravity, Urgency and Tendency of a particular issue to the problem	Establishes clear criteria to investigate a problem

Table 3.3: Steps of the scientific method and the quality tools that can be used to support its implementation in industrial research.

Step in the scientific method	Quality tools
Identification of the problem	Brainstorming Pareto diagram Flowchart
Definition of the problem	Mind maps Stratification Pareto diagram
Literature survey	Brainstorming (keywords)
Formulation of a hypothesis	Ishikawa diagram
Gathering and analysing data	Control graphs Histogram Mind maps Data sampling Correlation matrix
Formulation of a thesis	Ishikawa diagram
Publication	Mind maps

I will now demonstrate how you can implement each step of the scientific method using quality tools already discussed.

IDENTIFYING THE PROBLEM

Reference tools: brainstorming, Pareto diagram, GUT

In industry a problem is an undesirable result. This result can occur in an activity or in a process. It will not always be possible to solve the industry's problems; however, the effects can usually be minimised with minimal costs.

Frequently the complete solution to a given problem requires prohibitive expense. Therefore, it is necessary to carefully plan what actions you are going to take before looking for a good solution for a particular problem. To prioritise the problems under investigation, you should consider at least the following aspects:

1 How important is the solution of this problem for your organisation and its shareholders?

GUT and Pareto can help you here. Your time is precious; you need to use it well, and the more effective you are in using your time, the better it will be for your career, professional image and even your wellbeing. I can assure you that if you can identify relevant problems for your organisation and solve them effectively, you will advance in your career. However, some problems are easier to identify than others. Being efficient in finding relevant problems to investigate depends also on the organisation's culture. Some problems are very evident to new employees because they come to the organisation with 'new eyes'. You don't necessarily have to solve the problem completely, merely minimise its effects. Being efficient in identifying a problem, collecting more information and formulating clear hypotheses on the causes of the problem is already a great contribution in industry.

2 How long will it take and how much will it cost to get the results?

Anyone can come up with five problems that are absolutely relevant to their organisation or its clients! To be frank, you cannot solve them all at once. If possible, your problem in research should be a unique one. You should focus on deciding on the problem to be investigated and you should try to work on one problem at a time. You can use GUT to support you in the process of choosing which problem to investigate. In industry it is important to tackle and obtain results in the short term. Problems with potential outcomes that require a long time to be implemented should be considered again in the future. This is even more important if your department has not been producing good results for a while. Best practice, in fact, is to discuss all the problems to be solved and choose one of them to be solved first. If it is up to you to decide, in a heartbeat, take the one that will get the fastest results. And the other four problems? Let the future take care of them.

3 How much will the solution cost and what is the real benefit obtained with the (partial or total) solution to the problem?

Research is an economic activity with considerable risks and potential attractive profits. To run a profitable business in research, it is necessary to have good planning and efficiency. You can minimise risk by investing your time and resources in planning your research properly. You should consider research as any other economic activity and, as in an industrial business, you should have a production plan (e.g. number of patents, reports, projects with clients, revenues, new products, processes or services, publications and so on) to work on.

You, as a researcher, should ask yourself if you are earning your salary with your research outcomes. If you invest your time and your company's resources for a reasonable time in a project without any good result or impact, you should be worried. It can be understood as incompetence, lack of responsibility with the industry's resources and bad for your career.

LITERATURE REVIEW

Reference tools: brainstorming, Pareto diagram, GUT

If you want to solve a problem you must first understand it. In industry very few problems have been solved by chance.

The scientific method recommends that you search for the best available information about your problem. No one should expect to develop a good research project without carrying out a decent literature review. Research in industry, as a problem-solving activity, will involve some adoption of old ideas or techniques. This adoption process is worth doing. Results already published in the literature should be compared to your results in terms of value, precision and representativeness. On the other hand, you can have new ideas, or more resources than those earlier researchers had to develop their work. Therefore, you might be able to solve problems they were not able to solve. While doing a literature review you will become familiar with other researchers in the field and know their approach to similar problems. What is really intriguing, and what worries me, is to see many researchers not doing a good literature survey before commencing the work, not being able to do it as a useful or purpose-oriented process, or simply, and worse, replacing a literature survey with a 15-minute brainstorming session. Brainstorming will never substitute for a bibliographical review, and it only produces a list of keywords. In the days of internet and digital media, it is inexcusable not to complete a good literature review.

I am sure your organisation will have an extensive list of reports and publications available on the problem you are planning to solve. Otherwise, there will be, if not a vast, at least some literature available. A few minutes on any search engine, ISI Web of Knowledge or similar tools will give you a glimpse of what these services can offer in terms of a literature search. At worst, if the chosen problem is completely new to you, your client and their experience in the field you are working on will be a good starting point and there will be something published they can share with you. The point is – there is no excuse not to do your job. Resources are available and reading these is vitally important to the success of your project. Reading what was done about similar problems will help you to understand your problem better. By 'understanding better' I mean knowing what the common causes are and being able to identify the unexpected ones, the alternative methodologies applied to solve them, new interpretation of experimental data and new bibliographical references. These are just a few among many other reasons to learn from other people's experience.

Performing a good literature review is one of the most important aspects of the scientific method. I'll recap some of the reasons why.

1 To know more about your problem

Good papers and reports introduce similar problems and motivations for solving a problem from a well-structured perspective, and quite frequently these points of view are unknown to you. Doing a literature review and learning other people's perspectives will make it easier for you to justify further investment in your project, or even justify its existence.

The results obtained by others may shed light on doubts you might have about the extension of your problem. Can you imagine the difference it would make to your investigation if you discovered that climate change is forecast to be more intense over the next years and have a generally positive influence on agriculture in Tasmania?

2 To know about new experimental approaches

You may be used to solving your problems with a certain set of techniques and methods. The literature review may reveal new techniques of data analysis, sampling and material characterisation that may be well suited to your study. You have to accept that all your training and experience have not covered everything ever devised to solve problems! Ultimately, all good researchers understand that their formal education and years of practice result in a more efficient way of learning from others and developing their research from what was done before.

3 Who cited and who was cited

The literature review might provide you with new keywords to continue your literature search. The literature cited by other researchers may also be relevant to your work. Some of the tools available for the literature review have changed the way researchers undertake a literature survey today. Before the internet it was almost impossible to keep track of, for example, who cited the paper you happen to have in your hands. It was relatively easy to track papers cited by an author: those were listed in their bibliography. But it used to be almost impossible to see who may have eventually cited the paper you are holding. Today it is possible to know who was cited in a paper and who cited that same paper later on. When you access a paper using, for example, Web of Knowledge, you can access a list of references cited by the authors as well as a list of papers that have cited the paper you are reading.

4 To know who is working on a similar problem

Researchers are usually very focused in their area of activity. Few are likely to explore different disciplines. I encourage you to look for good authors in your field with whom you can exchange experiences: visit them, work with their students … very soon you will be working and publishing together. Collaboration in research is very common and should be encouraged.

The majority of researchers promptly answer requests from other professionals. Pay attention to your scientific and professional citizenship: be an active member of associations in your area. Some good examples are the Institute for Electrical and Electronics Engineers (IEEE), American Geophysical Union (AGU), Australian Institute of Physics (AIP) and the American Association for Advancement of Science (AAAS). These associations have specific publications, journals, conferences, white papers, manuals, catalogues of products and publications, electronic lists of specific discussions, awards and so on.

As mentioned before, the objectives of the literature review are to identify what has been done before about a given subject, the points of agreement among researchers and points of disagreement, as well as introducing you to new ideas about experiments that would test your hypothesis more effectively.

Remember: a literature review is not a list of published work, but a critical analysis of what has been done about a given subject and the identification of opportunities that can be explored in your research.

FORMULATING HYPOTHESES

Reference tools: brainstorming, Pareto, GUT

The formulation of a hypothesis requires some ability with data analysis and synthesis of concepts. At this stage you may have decided on a very specific problem to investigate. You should also have done a literature review where you:

- read scientific papers on the same subject
- read equipment manuals
- held interviews or had less formal discussions with other specialists in the area
- collected standards, and
- collected other relevant documents or even collected and analysed some data from your reviewing process.

All this effort has a very specific objective: to better understand what you are going to investigate. This should lead you to a crucial point in your research: the formulation of a winning hypothesis.

The hypothesis is the synthesis of your understanding on why and how a problem occurs. At this stage your hypothesis still needs to be tested. Remember that in science the only accepted hypotheses are those that can be tested.

Those working with problems in industry (either products or services) should look for a hypothesis that can be tested quickly and which will not be excessively costly. The cost/benefit ratio should be low. Imagine you are asked to solve the problem of high fuel consumption data in a fleet of trucks. You have examined some data on fuel consumption from the company, considered the load the trucks are carrying, interviewed drivers, read the reports on maintenance and inspection, have been on some of the trips (to inspect the quality of roads and their profile), spoken with engineers working with the truck suppliers, compared data with other companies you have full access to ...

After all this (the literature review), you have analysed the data, compared data from drivers, distances and truck suppliers – you should establish a hypothesis. No matter what hypothesis you propose, you should propose a solution that is easy to test and implement. As with any scientific research, you should test your hypothesis. The proposed solution should be accompanied by an experimental plan to verify your hypothesis. Perhaps you can establish a cost reduction target for your hypothesis. This target, guaranteed by your tested hypothesis and by implementing it properly, could be used to define cost reductions. But what will your proposed solution be? There is a long list of possibilities (that a brainstorming session may have produced): new trucks, replacement of some parts, a change in the supplier of fuel and oils, training provided to drivers, new alternatives in logistics, a combination of these … the solutions are up to you. Bear in mind that an industry solution needs to be simple, easy to implement and produce a high return on investment. For example, if your conclusion is that the fleet is not appropriate for the size of the load, you need to propose, if possible, changing the size of the load and not replacing the fleet!

Another important piece of advice: organisations are made up of professionals who are very experienced, assured and confident about what they do. This is true for almost all industries. You will hear the following kinds of comments from them:

'I studied this problem many years ago and the solution is to do … or …'

'I'm sure the system is fine the way it is!'

'There is a lot of data that proves this variable is basically constant.'

'Mate, I wrote a report about this in 1975 …'

The first way to deal with this confronting situation is to ask to see the professional's technical report on their solution (incidentally, this is what you should have done while carrying out your literature review!) and their proof that the 'variable is constant'. The point is: ask for data and facts. In most cases, their prior assumptions will not be grounded in observed data and facts. You can always argue that you 'believe' he or she is correct, but you need to prove it with new data and facts. Belief in the absence of compelling evidence is faith. By the way: if there was a conclusive study in 1975, why is there still a problem?

Personal assurance – when established by a group of people it is known as 'common sense' – can result in a misleading hypothesis.

Common sense is the result of personality and experience, and can often come up in brainstorming, in the interviews you conduct or even in technical reports. You need to be alert and not have your work and hypothesis misled by common sense. Experienced professionals are usually great project managers. Try to have them on your side, but don't let them influence your work with their 'beliefs'. If the assured professional is your chief and this is a big problem for you, I recommend you go to the closest library and read a good text from the self-help section!

SAMPLING, ANALYSIS AND DATA INTERPRETATION

Experiment planning (sampling)

Many professionals working in industries (and universities!) carry out experiments without any idea about what they are expecting to obtain. Very often they do them because they are interesting. This terrible practice leads to loss of time, effort and money. Professionals can, if they indulge in such a practice, lose their credibility (and any credibility in their good methods) because they are doing interesting things, but not solving any problems. They should be using tools and not toys. To avoid this situation, some laboratories use fines to discipline internal clients (other departments or students). This practice encourages thought and planning before the researchers get to work. Doing without planning is foreign to the scientific method and to the responsible researcher. It is always necessary to plan your experiments and 'guess' what you might obtain from them.

Planning experiments requires experience. Experience comes with time and working with good people, but there is a short-cut: the literature review. One of the most delightful exercises of research is the sense of insight you get from reading the literature.

A literature review, as I said in the previous chapter, is the opportunity to capture ideas that have not been tested yet. If the literature presents a good result obtained from a similar problem, repeat the solution and get the problem solved that way. It may also help you to not waste time with 'common sense', a result somebody has proved to be wrong, or it may help you to propose a hypothesis nobody has tested before. The latter case may lead to real innovation in industrial research.

The important fact here is that for similar problems already solved, the experiment planning should be almost ready. If your hypothesis is original there is good potential to capture many research outputs (e.g. from scientific papers to patents and other IP instruments). In this case, you need to plan your experiment with care and document everything as well as possible.

Because of the nature of innovative projects, it is likely that a few experiments were not planned at the time you established the hypothesis and set out the required experiments. This requires projects to be reviewed 'on-the-fly'.

The key aspect to any experiment is there must be a reason for collecting data. You cannot collect data without planning. Collecting historical or fresh data without a good reason is a waste of time.

Data analysis and interpretation

Data collection is the systematic work of recording entities, taking into consideration uncertainties involved in the measurement process. Scientists apply different tools to find patterns and trends in data sets. Other tools can help interpret the phenomena they are studying.

The quality of the interpretation is the key to innovation in research. To reiterate, development is about hitting the target presented to you; innovation is hitting the target nobody sees. Interpretation, aside from what can be based on very specific theories, is extremely personal.

Interpretation, in a chess game, is what differentiates a normal player from a Grand Master.

There are a few roadblocks to 'correct' interpretation of data and phenomena. These can be self-assurance, common sense or even an immediate association. Others are procrastination, poor experiment planning, poor data gathering, trusting in data without 'questioning the numbers you see' and poor background information.

For example, it is very easy to find a clear pattern on people's weight and how they manage their diet. Light and diet drinks are usually consumed by people trying to lose weight or by those with diabetes. Just looking at the data you could imagine that consumption of light and diet drinks causes obesity and diabetes because people who don't have these problems usually do not consume them!

This is an extremely common situation in research. Analysing data without thinking can lead to false conclusions. Innovation comes from a new way of thinking; the result of rigorous intellectual effort is an innovative interpretation of data and phenomena. The key to success in innovative research is careful thinking.

SOLVING PROBLEMS IN RESEARCH

There are some common problems in research. In most cases they occur because scientists are negligent in some of their responsibilities. It is important to identify and react when the first signs of problems appear.

Here is a list of problems scientists stumble upon when performing research:

- problem selected is too broad
- inability to meet deadlines
- shortages of resources before the research comes to a conclusion
- poor literature review
- lack of experiment planning
- poor data-gathering process
- inadequate data analysis
- lack of originality
- poor interpretation of experimental results
- poor presentation of research results
- excessive assertiveness or confidence and insufficient arguments
- not being honest with allocation of credit to the work
- duplication
- falsification
- fabrication.

One last message: in industrial and academic research, poor implementation is less expensive than poor planning. Remember this before you start your next research project.

Chapter 4

The quality of your measurement

PLANNING EXPERIMENTS

Have you ever considered that the sampling method you use in your experiment can affect your conclusions?

Experiments are meant to be well planned and purpose-oriented (i.e. designed to verify your hypothesis). It is important to plan your experiments and data sampling before you start measuring anything. Experiment planning is the architecture and the strategy of your hypothesis.

In general, all systems or processes are made of four sets of data:

- input data
- controllable variables
- uncontrollable variables
- output data.

Figure 4.1 illustrates this typical model of processes and systems containing input and output data, as well as controllable and uncontrollable variables.

Uncontrollable variables might include things such as the room temperature, vibrations in the laboratory, variations in power supply to the electronic balance, sound, and other external events that may affect your experiment. It is possible to control them completely, but the cost can sometimes be prohibitive. The best way to deal with them is to

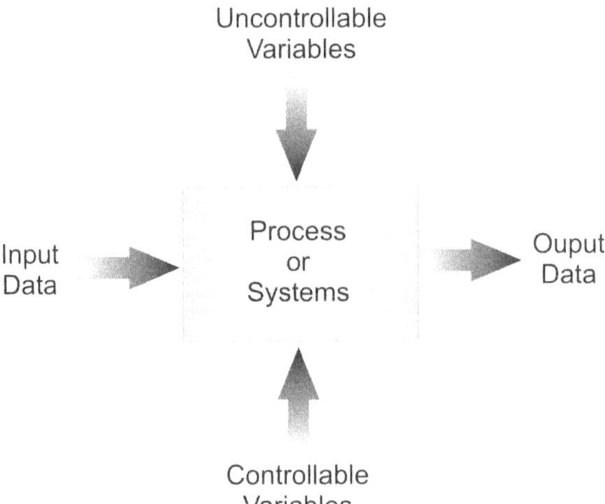

Figure 4.1: Data and variable types in a general data model of processes and systems.

repeat experiments to capture their variation and express them as an uncertainty of your measurement procedure.

A controllable variable could be the set-up temperature of a furnace, the pH of a solution, the relative pressure of a gas, the colour of a pigment or the weight of a sample.

Industrial experiments are usually deployed to study the performance of processes. Industrial processes can have complex combinations of machinery, raw materials, operators, working conditions and methods. Variables associated with these processes, once quantified, will be your input data. Some variables will be controlled in your experiment and others may be uncontrollable, but at least they should be recorded in your experiment. This record could be extremely important to explain the limits of your experiment or to help you explain deviations from expected behaviour.

Experimental planning should include at least the following steps.

- Determine which variable most influences the output data. As seen in the previous chapter, you should have some quality tools to support you here.
- Determine the range of values the controllable variables will have, so that the uncontrollable variables are left to vary as little

as possible. The effects of uncontrollable variables in experiments should always be small – perhaps better experiment design could help here.

Strategy for experiments

On 14 January 2004, a few hours before receiving the first data from the miniaturised spectrometer installed in the rover working on Mars, the scientists involved with that experiment made a bet: what would the shape of the spectrum be and why?

It was interesting to me to see that the specialists working on that mission had quite diverse ideas about the shape. And, because spectral shape is related to minerals and their properties, scientists differed in their hypotheses. Some of these were incompatible. The scientists revised their guesses again before new data came through.

The scientists on the Mars project were involved in a most interesting exercise. Not only were their competitive instincts aroused, but the days following the rovers on the Red Planet were spent elaborating on hypotheses and arguing on the type of operations that should be carried out. As the mission developed, more and more data came in to support the scientific rationale of doing experiments on Mars.[1]

As of today the exercise continues, after more than six terrestrial years of daily operations carried out by the Mars rovers. While the scientists' hypotheses were in place before the rovers landed, they were tested, improved and changed as data accumulated, and this has resulted in significant changes to the body of knowledge regarding the existence of water on Mars.

The important fact is: if you can foresee the results of a measurement, you may be able to plan your experiment more effectively.

Before starting on your experiments you should draw up a detailed plan. Lack of planning, especially in industrial research, may lead to data that is not interpretable. Poorly planned experiments frequently lead only to a waste of money. If you are involved in field experiments, you should have at least a collection of spare parts, communication devices and other support for a contingency plan.

Bear in mind the limits of your observations, the precision of the applied instruments, the method to be used, and again, what you expect to obtain from your measurements. A list of references on experimental planning is provided in the endnotes.

WHY SHOULD YOU TREAT YOUR DATA CAREFULLY?

No matter what experiments you perform, you should understand the degree of precision you are aiming for, the uncertainties involved, the expected degree of repeatability and the reproducibility of your experiment. I will now discuss these concepts a bit further and illustrate them with simple examples.

Errors are usually classified into two groups: systematic and statistical errors. Accuracy is associated with systematic errors. Precision is associated with statistical errors. Therefore, instruments with good accuracy have small systematic errors, and instruments with poor precision have large statistical errors.

The idea of data quality

The quality of the measurements obtained with your data needs to be reported with care. This means considering elements such as bias, precision, accuracy, repeatability and reproducibility. Associated with a measurement method there is a statistical concept of distribution. I will visit these concepts and their implications for the analysis of your experimental data.

It is a vast subject and there are excellent texts available. I would like to focus on some relevant, basic concepts you should consider while performing your experiments.

Distribution

The distribution represents the statistical model that captures the typical variation obtained with the use of a given measurement method. The distribution is the expression of all measurements of a given entity with a given instrument under a given condition. But if I measure the same thing, with the same instrument under the same conditions, why don't I get the same result?

The answer is simple: whatever you measure changes with temperature, it is not homogeneous and it is not a constant. The instrument you use to measure with is not absolutely precise and is not applied the very same way every time you use it. The conditions are not always the same: temperature, for example, may change during the measurement, affecting in different ways both what you measure and the equipment you use. And you may have no control over these variations. Additionally, the act of measurement itself may affect the thing that is being measured.

Discrepancy

When two measurements of the same quantity disagree, there is said to be a discrepancy. The difference of these measurements will give you the discrepancy in quantitative terms.

Precision

The precision of an instrument determines how close independent measurements obtained with this particular equipment are. The smaller the variation obtained from independent measurements, the higher the precision of the applied measurement procedure.

Accuracy

The accuracy of an instrument determines how close the results obtained with this instrument are to the real value. Instruments with low deviation (bias) and high precision will have good accuracy. Any other combination of deviation and precision will result in low accuracy. Therefore, high accuracy will only be obtained from precise and unbiased equipment.

Repeatability

Repeatability of equipment is the variation obtained when an operator uses the instrument multiple times to determine the value of what is being measured. Instruments with poor repeatability produce wider distributions, while instruments with good repeatability produce sharp distributions. You can use the standard deviation to express repeatability of a single instrument, but if you want to compare the repeatability of different instruments used in a single experiment[2] you should use the coefficient of variation. The coefficient of variation is the ratio of the standard deviation to the absolute value of the mean. It expresses how many standard deviations you have in one mean.

For normal distributions, 99.7% of the measurements will be contained within three standard deviations around the mean (mean ± 3 standard deviations).

Reproducibility

Reproducibility of an instrument is the variation of many measurements, especially if performed by different operators. This is a very important concept in science because experiments developed (and published) by one researcher should be reproducible by other researchers.

Repeatability and reproducibility are associated with the precision of an instrument or measurement procedure.

The total deviation of a set of measurements depends on two aspects:

- variability of the measurement procedure
- precision of the instruments applied in the measurement.

It is possible to estimate the contribution of the measurement procedure and the instruments.

To estimate the repeatability and the reproducibility of measurement systems of samples or units produced in a factory:

- randomly select the pieces for measurement
- samples selected for measurement should cover the entire range of specification
- select a reasonably large number of samples for this assessment
- a good practice is to repeat the measurement on some pieces already measured
- before moving on to a new piece, consider if the result you obtained makes sense
- verify if the uncertainty of your measurement really does make sense for the applied method. For example, can you obtain precision in micrometres using a common ruler?

The subject I am trying to cover here could fill an entire book and I have listed a few books in the endnotes that I recommend for further reading. However, I have decided to discuss these concepts here because I feel that they will aid you to understand what is written in scientific papers and you have to know them if you want to perform your own experiments with the care and attention that is necessary.

Next I'm going to introduce a few useful statistical tools and illustrate how they can be applied with some examples. Again, a list of worthwhile books is provided.

SOME USEFUL STATISTICAL TOOLS

In most of the experimental work you perform you will report the results you obtain using numbers. Once processed, this data will provide you with the information you need to use to challenge or verify your hypothesis. However, the measurements you take may be subject to errors you cannot eliminate. No matter how experienced and careful you are, errors and deviations will usually be present. You cannot guarantee you obtained the true value of something you measured with a single measurement. Measurements are expected to carry information such as, for example, how close successive inferences are to the real value and with what certainty you can assert that your measurement is close to the real value. As you can see, you should expect to express the measurement with two pieces of information: what the expected value is, and how sure you are about it.

This is not very intuitive because in daily life people are used to providing a single piece of information: how far, heavy, cold and so on.

You need to identify the possible source and sort of errors associated with your measurement procedure. I've discussed systematic and statistical errors earlier in this chapter.

Poorly calibrated instruments, with defective construction, or inadequately used instruments may produce systematic errors. For example, if you use a thermometer with the scale shifted from the appropriate position this will cause systematic errors. Or if you read the voltage using multimeters without their scale being calibrated at the beginning, you will also get results with systematic errors. Researchers can reduce systematic errors through using appropriate and accurate instruments.

Statistical errors are caused by uncontrolled and random variations of equipment (for example caused by variation in power supply or electronics temperature) or of the environment (for example humidity, light, room temperature).

Now I'll ask you a simple question: Is the equation below correct or not?

$$2 = 2.0$$

The answer is: quantitatively it is correct, but qualitatively it is incorrect.

If you measure the temperature of a gas and you register 300°C, you are not saying the temperature is exactly 300.00000...°C, but that the temperature should be close to 300°C.

It might be reasonable to expect that the temperature is smaller than 310°C and larger than 290°C. Note that there is always an uncertainty associated with an experimental measurement. This uncertainty is usually half of the smallest scale unit you have in your measurement device. For example, if the smallest scale you have is 0.1°C, the temperature might be assumed to be between 299.05°C and 300.05°C.

Similarly, if you read the diameter of a coin using a ruler with millimetre scale and you obtain 10 mm, you might expect that the diameter of this coin is between 9.5 mm and 10.5 mm. Note that the number of significant digits are associated with the accuracy of the instruments you use.

Accordingly, the quality of a measurement will be associated with how careful you were while calibrating the instruments, how much effort you put into eliminating possible systematic errors, your understanding of possible sources of statistical errors, the correct data reading, and the independent measurements of a long series of data. If you are working in the research domain and have never paid attention to these details, you are likely to have made many mistakes in the interpretation of your results. An assessment process without addressing these issues is not a measurement, but speculation.

Research is about being careful in readings, planning before taking measurements, considering the right instrument for each measurement, and being extremely careful with the quality of your data. A good and experienced researcher does this effortlessly. You cannot be a successful researcher if you do experiments without taking significant care. The situation becomes more serious when an industrial researcher needs to use data recorded by others. In this situation the researcher needs to verify how the data was measured. I am shocked when I see scientists working with pieces of information, using data without questioning data provenance.

As I said before, no measurement is error-free, no matter how careful you are. There will always be errors. The question you have to ask is: how big is the error associated with your measurement?

To estimate the error associated with your measurement, you should repeat the measurement several times. The differences between many independent data sets you obtain will provide you with an idea of your experimental error. Independence between these measurements is necessary. Otherwise one measurement will interfere with the other. Therefore, it is important to identify the best procedure to ensure that systematic errors are smaller than statistical errors. Otherwise, you will introduce errors while trying to estimate them!

For example, imagine that every time you measure the volume of liquid mercury at approximately 20°C that you are heating the container with your hands. The mercury expands much more than the glass of your container. Hence, the apparent volume of the mercury will increase because you are heating it with your measurement procedure. In this case each measurement is interfering with the other and your measurements cannot be considered independent.

If the liquid mercury was at the same temperature as the container and you did not interfere with its temperature with successive measurements, you can say your measurements were independent.

The measurement will approach the true value when its variation is as small as the statistical errors associated with successive measurements. For example, if you measure the diameter of a coin using callipers you may obtain different results because you cannot always guarantee you will be measuring the diameter of the coin at some points. In this situation you can say the instrument is suitable enough for your measurements. On the other hand, when the variation of your measurements is bigger than the accuracy of your equipment, the differences can reveal the variation of what you measure. Statistical variations like these are what you want to capture with your experiment.

Two statistical concepts underlie what researchers are trying to measure with their experiments. They are called measurement of central tendency, and dispersion. The central tendency tells us the value around which the measurements are distributed. Your experimental procedure will be appropriate if the dispersion is small compared to the precision of your equipment. Dispersion represents how widely distributed your measurements are.

Measurement of central tendency

I will now introduce six different measurements of central tendency. They are the arithmetic mean, the geometric mean, the harmonic mean, the weighted mean, the median and the mode.

Arithmetic mean

This is the simplest and by far the commonest measurement of central tendency. It is expressed as the average value of successive measurements. For example, if you measured the acidity (pH) of a solution seven times and you obtained 2.1; 2.5; 1.9; 2.4; 2.1; 2.3 and 2.1, the mean value (a) will be the sum of your seven measurements divided by seven.

$$a = (2.1 + 2.5 + 1.9 + 2.4 + 2.1 + 2.3 + 2.1\ ) \,/\, 7 = 15.4 \,/\, 7 = 2.2$$

Geometric mean

Geometric mean (g) is obtained with the nth root of the product of all n measurements. For the example with pH measurement we would have:

$$g = (2.1 \times 2.5 \times 1.9 \times 2.4 \times 2.1 \times 2.3 \times 2.1)^{1/7} = (242.82342)^{1/7} = 2.19$$

The geometric mean is extremely useful for excluding a possible wrong measurement – one that deviates substantially from the others. For example, if one measurement is very close to zero, this might mean two things: your liquid is extremely acidic (probably unexpected in your measurement and you should therefore explain this) or the measurement is wrong (measure that liquid again). On the other hand, the arithmetic mean will ‘hide’ results like this. Another interesting application of the geometric mean is when you use subjective criteria to decide upon different options. Take as an example the quality of wine influenced by climate change. To work on this problem you need to define what you mean by quality. For example, assign a rating from 0 to 10 for each of the following criteria: scent, colour and taste. You can use the geometric mean from these three criteria to decide which wine is

Table 4.1: Grades given to each year for a given wine, with arithmetic and geometric means presented.

Wine	Scent	Colour	Taste	Arithmetic mean	Geometric mean
2004	7	7	8	7.33	7.32
2005	6	7	9	7.33	7.23
2006	8	4	10	7.33	6.84

the best. Table 4.1 presents hypothetical grades to each of these quality parameters of the wines an expert tested.

From Table 4.1 you can conclude that the arithmetic mean suggests that all wines are the same. But you could conclude that the best wine (or most balanced wine) is the one with the highest geometric mean (year 2004).

You may argue that these characteristics should not be considered equally. A specialist might say taste is more important than colour and colour than scent. To consider this we should use the weighted mean.

Weighted mean

If taste is twice as important as scent, and colour is 50% more important than scent, Table 4.2 presents the result given these criteria.

In accordance with these new criteria the best year is 2005.

Imagine that you mix the liquids contained in seven different containers. Imagine that the pH and the volume of each container are different. Table 4.3 presents values of pH and volume for each container.

Calculating the weighted mean to obtain the pH that will result after all liquids are mixed together is given in the following equation.

Table 4.2: The evaluation of wines using weighted criteria.

Year	Scent (weight 1)	Colour (weight 1.5)	Taste (weight 2)	Weighted mean
2004	7	7	8	$(7 \times 1 + 7 \times 1.5 + 8 \times 2)/4.5 = 7.44$
2005	6	7	9	$(6 \times 1 + 7 \times 1.5 + 9 \times 2)/4.5 = 7.67$
2006	8	4	10	$(8 \times 1 + 4 \times 1.5 \times 10 \times 2)/4.5 = 7.56$

Table 4.3: Volume and pH of each container.

Container	pH	Volume (mL)
1	2.2	300
2	2.3	210
3	1.8	170
4	2.4	190
5	2.6	280
6	2.0	130
7	2.1	260
	Total volume	1540

$$\text{Final pH} = (2.2 \times 300 + 2.3 \times 210 + 1.8 \times 170 + 2.4 \times 190 + 2.6 \times 280 + 2.0 \times 130 + 2.1 \times 260)/(1540) = 3439/1540 = 2.233$$

But because your precision is one decimal digit you should round the result from 2.233 to 2.2.

Harmonic mean

In certain situations, especially those involving rates and ratios, the harmonic mean (h) may provide the truest average. For example, the average of a collection of electric resistors connected in parallel arrangement in a circuit is given by the harmonic mean.

$$1/h = 1/n\ (1/n_1 + 1/n_2 + \ldots + 1/n_3)$$

The harmonic mean for the liquids with different pH will be:

$$1/h = 1/7\ [1/(2.1) + 1/(2.5) + 1/(1.9) + 1/(2.4) + 1/(2.1) + 1/(2.3) + 1/(2.1)] = 1/7\ [3.206]$$

Therefore,

$$1/h = 0.45804$$

And the harmonic mean will be:

$$h = 2.18$$

It is possible to demonstrate that $a \leq g \leq h$. The arithmetic mean, geometric mean and harmonic mean will be the same if all individual values were the same.

Median

Median is the value that is in the central position of an ordered set of data. If the number of data readings is odd, the value is promptly available; if even, the value is obtained by the arithmetic mean of the pair of numbers in the central position.

For the pH of the liquids you obtained the results in the following sequence:

2.1; 2.5; 1.9; 2.4; 2.1; 2.3 and 2.1

Ordered they will result in:

1.9; 2.1; 2.1; **2.1**; 2.3; 2.4 and 2.5

The median is indicated by the central number (that is 2.1).

Mode

Mode is the most frequent value found in a sequence of data. For the pH the most frequent data is 2.1. It appears three times in the series of measurements. On the other hand if no value repeats, the set of data has no mode. A data set can, however, have more than one mode if values you read repeat at least once and each group has the same number of values.

There is a vast amount of literature available about basic statistics. I've touched upon the topic in the book because you cannot be innovative in research if you don't consider the quality of your data, if you don't use the appropriate tools to analyse your data, or if you don't understand the relevance of minimum experimental care while collecting or using someone's data.

As stated earlier, by repeating measurements taking all necessary care you will obtain results clustered around a value (that is the central tendency). I will briefly discuss some assessment criteria on the dispersion of data.

Measurements of data dispersion

In the previous section a few measurements of central tendency were discussed. These calculations are performed to find a value towards which all data converges. Measurements of central tendency are not appropriate for expressing how dispersed your measurements are. You need other tools to do that. I will discuss four of them to provide you with some understanding of data dispersion: range, variance, standard deviation, and coefficient of variation.

Range

Range is the simplest measurement of dispersion. It is the difference between the smallest and the largest values you obtained with your measurement. For example, for the pH measurements the range of dispersion is given by:

$$A = \text{higher value} - \text{small value} = 2.5 - 1.9 = 0.6$$

Variance

Note that to calculate the range you only use two data samples from the entire series. No other data is used. By calculating range you are not using all the available data and, therefore, you are not capturing all the information your data can reveal.

To overcome this limitation you need to use another dispersion measurement. Because data is dispersed around a central tendency value, we may try to measure how data is distributed around, for example, the arithmetic mean (herein mean). This distance can be called the deviation related to the mean. If you add all deviations you will get zero. But if you consider the square of the deviation you will arrive at the distribution. You might have 10, 1000 or any other number of measurements in your series. The number of samples is relevant when you calculate the distribution around the mean. Therefore, another important aspect is the number of samples you have in the series. The number of terms minus one is known as the degree of freedom of the system. You can define variance as the sum of the deviations squared, divided by the number of terms minus 1.

To explore this definition a little bit further, look at the example below – the pH measurements:

2.1; 2.5; 1.9; 2.4; 2.1; 2.3 and 2.1

Because there are seven measurements, the degree of freedom of the system is six. Table 4.4 illustrates step by step the calculation necessary to obtain the variance.

Table 4.4: Calculation of the variance associated with the pH measurements.

Result of each measurement	Measurement minus the mean	Deviation square
1.9	1.9 – 2.2 = –0.3	$(-0.3)^2 = 0.09$
2.1	2.1 – 2.2 = –0.1	$(-0.1)^2 = 0.01$
2.1	2.1 – 2.2 = –0.1	$(-0.1)^2 = 0.01$
2.1	2.1 – 2.2 = –0.1	$(-0.1)^2 = 0.01$
2.3	2.3 – 2.2 = 0.1	$(0.1)^2 = 0.01$
2.4	2.4 – 2.2 = 0.2	$(0.2)^2 = 0.04$
2.5	2.5 – 2.2 = 0.3	$(0.3)^2 = 0.09$

The sum of all deviations squared is 0.26.
The variance (s^2) will be:

$$s^2 = 0.26/6 = 0.043$$

The standard deviation

Variance is a measure of dispersion. It considers all data you have, the mean, and even the number of samples there are. But there is a limitation: the value is somehow meaningless: the unit of variance is the square of what you really measure. If you measure the distance in metres, the resultant unit will be square metres because you've calculated the square of the differences. If you measure temperature in kelvin, the variance will be expressed in K^2. It is difficult to understand how dispersed your data is if the unit is always the square of what you really measure. It would be better if you could measure the dispersion using the same unit as you are doing the measurement in. The positive square root of the variance, which has the same unit as the entity you assess, is the 'standard deviation'. Because standard deviation and mean are in the same unit, interpretation of the data is more intuitive.

The standard deviation of the pH measurements will be:

$$\text{s.d.} = 0.21$$

You could express the value resulting from successive (and independent) measurements of pH as the mean ± three standard deviations

(2.2 ± 0.6). In this case you would expect to have approximately 99% of the measurements in this range: from 1.6 to 2.8.

The use of mean and standard deviation to estimate the confidence interval in natural systems is beyond the scope of this book. You can read some of the excellent references set out in the endnotes.

Note that the standard deviation (s.d. = 0.21) was expressed with more significant figures than when you express it as an uncertainty of an experiment (0.6). It is good practice to express the last significant digit in any uncertainty expression with the same order of magnitude as the measurement. An exception would be when the uncertainty is 1 and the mean value is 2.7. You may want to express it as 3 ± 1. Note that the answer of 2.7 ± 1 is quite acceptable because rounding it would waste information from the mean.

Coefficient of variation

It is very difficult to compare the accuracy of two or more series of independent measurements. For example, consider an experiment where you measure temperature, volume and pressure. Imagine that you have measured these parameters 100 times, and calculated their mean and standard deviations. The key question is: which series has better accuracy given the results below?

Temperature: (298.2 ± 0.4) K, where 0.4 is one s.d.
Volume: (2.25 ± 0.15) L
Pressure: (300 ± 3) hPa

To know which parameter has been measured more accurately, you can use the coefficient of variation as the ratio of standard deviation and the mean multiplied by 100. The result will be expressed as a percentage. For the data above the result will be:

Temperature: (0.4/298.2) × 100 = 0.13%
Volume: (0.15 / 2.25) × 100 = 6.7%
Pressure: (3 / 300) × 100 = 1%

The coefficient of variance for each measurement above reveals that the temperature has been measured more accurately than the volume. If in your experiment you use these three measurements and you need to improve the quality of what you measure, you can easily see that you need to improve the way you measure volume.

CONCLUSION

Data in industrial research are the key element in your investigation. Whether you use data provided by others or data collected yourself, you should be extremely careful: wrong conclusions can be drawn from poor data.

Industrial data is generally poor in quality. You may need to analyse the data, interpolate, extrapolate, change the frequency to have all data on the same basis and, not surprisingly, you will have to assess the uncertainty involved in the experimental procedures.

Working in industry you will note that many tables, spreadsheets or databases have no expression of uncertainty. In this case, it would be entirely up to the reader to guess how precisely data has been measured – this is most unsatisfactory.

While designing your experiment you must have a clear understanding of your physical model, create a mathematical representation of this model, select a measuring methodology, measure, evaluate, and produce an estimate and an uncertainty associated with your experiment. Ideally the measurement uncertainty around the estimate should contain the true value of what you are trying to measure.

You can learn so much from your data using simple tools. One important lesson could be how powerful a measurement like coefficient of variation can be.

The objective of this chapter was to present a basic statistical analysis, also known as descriptive statistics. This analysis is the most common technique applied to experimental data. Unfortunately it is not common in industrial practice to report the uncertainty values or standard deviation of a set of measurements. I would argue that expressing uncertainty provides a basis for further work on improving the measurement.

Chapter 5
Research management

In this chapter I will discuss some important aspects of managing research projects in private or public organisations. Research projects have some unique characteristics and, therefore, unique needs.

One of the most important aspects of a research project is the team involved. Successful research managers are known to be great motivators. They are also willing to explore ways to get the research team to work well. Communication between technicians, engineers and scientists is not usually easy. Mentoring and coaching researchers is a difficult task. Building strong teams in research is far more challenging than in many other areas.

I will explore important aspects of collaboration in research projects, how interdisciplinary projects should be managed, how to work with distributed teams, and the risks and advantages you will face while working in large collaborative research projects.

THE RESEARCH PROJECT

I am not going to discuss the research project as a document scientists use to apply for grants. The research project in this book is a living activity that involves the management of teams around specific established objectives, strategies, deliverables and milestones.

As you can imagine, research activity can be flexible while researchers are defining the problem, performing a literature review and establishing hypotheses to be tested. However, the fact that researchers can start a research project with hypotheses to be defined cannot be an excuse for a relaxed project management. You cannot neglect the basics of project management in industrial research. As a researcher you should work hard to get the usual project management practices in place once the hypothesis is defined, as well as the experimental approach you are going to use to test it. Then it is a matter of executing the experiments you planned and making an honest analysis of the results. Alert researchers are able to capture details that were not expected in the experiments they originally planned. Innovation comes from sound intellectual process and, very frequently, from the careful analysis and good interpretation of something you were not expecting to see in your experiments. Does this fact influence the way you manage research projects? Certainly it does, but you cannot expect to make great discoveries in every single project you participate in. Research projects have to be treated as any other project (e.g. milestones should be established and delivered on time). When something unexpected happens you may consider changing the scope of the project. This is possible, but in doing so you should be careful. Discuss it with your stakeholders, present your scientific rationale and explain why it would be worthwhile to change the scope of the project. Then discuss the details of the new project to be implemented. Make sure all involved parties are aware of your plans to change the project (or abandon it). (Can you imagine what would happen if a builder decided to change the internal design of a house he is in the process of building without first discussing it with the owners?) Do you want to be successful in developing innovative solutions in industrial research? If so, make sure you get management on side for your research projects.

A research project can be considered like any other project in many aspects. The essentials are, therefore, the same. But research projects have some peculiarities you should be aware of when managing or working on one. Researchers, overwhelmingly, are very bad managers! They know there is always uncertainty in undertaking research, so they don't see much need for managing a research project. It is very common to see great emphasis in research institutes on a 'copy-and-paste' description of their science area that even extends to some of the claims regarding their research questions. Researchers are, in general, very conservative in committing to any outcomes of their work. However, they should not dissociate from reality and should be responsible with the public and private resources they manage. Many researchers in public and private research centres invest months of their lives reading papers and writing descriptions of what they read without any sense of applicability of what they are working on. All researchers should ask themselves if what they produce is worth their salary and the infrastructure they use. The bare minimum of project management skills would avoid these problems. In the following sections I will explore some aspects of research management.

Management of research projects is about setting up a clear research objective and supporting a research group to achieve excellence in their activity. A research team may include engineers, supporting staff, technicians and researchers. Any research project should contain:

- a clear statement of its purpose and scope
- the project's objective
- timeframe and milestones
- hypothesis, experimental strategy and expected results
- budget and assumptions
- expected output, outcomes and impact
- success criteria
- quality assessment
- educational and outreach activities
- risk assessment.

If you run a research project you should clearly set out all the above. But can a research project be successful without defining each one of these points? To answer this question, I will take success criteria as an example. If you don't define in advance what it is you are hoping to achieve, you can say any result is successful! A research project that does not have a risk assessment and well-documented lessons learnt is a bunch of people wasting money and not learning from their own mistakes. I consider it a contradiction of the fact that the scientific method contains the publication of results and a literature review – researchers are not gaining anything from the lessons learnt from projects within their organisation and scientific community.

If you work for a research project, ask yourself if you and the team working with you can define the abovementioned aspects. If you ask your team to define the objective of your project and they come up with different answers the team has no leadership in place, their project is poorly managed and the result will be a disaster.

THE RESEARCH TEAM

Having a good team is extremely important in projects. Research teams can be quite complex depending on the project characteristics. For example, I have worked in projects with two staff (myself and an engineer) as well as in projects with over 700 professionals. I have been involved in the implementation of projects worth less than $20 000 to more than 1 billion dollars. No matter the size of the projects you are involved in, you need a good research team to be successful. Your team has to work together to be successful – by nurturing their needs and expectations they will be highly productive. In research you can easily define three groups of professionals that have different needs and expectations. The groups are the researchers, the engineers and technicians, and the supporting staff.

Researchers

Researchers can be very motivated if they are working in their area of expertise and are working on a challenging project with potentially great outcomes. Researchers are usually proud of their work and, being hard workers, they can infect the entire team with their motivation.

Researchers can be extremely creative and motivated, they like recognition and some freedom to explore 'ideas'.

Things you can do to get researchers motivated:

- support their ideas or provide lots of constructive criticism to improve them
- provide them with resources to develop and explore new ideas
- recognise them for what they do
- challenge them
- celebrate each success. (These are less frequent in science than you may imagine.)

On the other hand, it is easy to destroy a scientist's motivation. You will do this if you take away their opportunities to do what they love or restrict them with small budgets, reporting routines, and poor recognition. Scientists lose motivation if you:

- don't support their ideas
- don't recognise or celebrate their successes
- don't provide them with resources
- ask them to do work that is very easy
- don't listen to them
- load them with bureaucratic work
- ask a bureaucrat to judge the value of their research.

Engineers and technicians

Professionals such as engineers and technicians are usually extremely realistic about their work.

To get engineers and technicians motivated:

- be bold, specific, and clear about what you are asking them to do
- give them an important piece of the project to be done
- share the benefits and recognition of the project with them
- give them a real sense of being part of the team.

To upset an engineer is easy:

- ask them to do something useless
- ask them to do something you do not define well
- frequently change your idea about something they are doing
- underestimate the workload associated with their task
- don't credit their work in publications
- swamp them with meetings.

Communication between scientists and engineers is not easy. Mostly it is because scientists are poorly committed to providing engineers with what they need: a clear activity to be responsible for and recognition for the success of this activity. Conflicts between scientists and engineers are therefore very common. While managing a research project, the leader should pay attention to the first sign of a problem and act promptly.

Supporting staff

Large research projects require a number of professionals who support the team with specific expertise. These can comprise media specialists, communicators, a secretary, security staff, drivers, divers, pilots, lawyers, Human Relations and Health, Safety and Environment professionals. Some of them will be absolutely essential to the success of your project.

Ensure the team is engaged early on in the project activities, define performance criteria for regular assessment, and give them frequent feedback. There is a simple principle to dealing with research team performance and motivation issues: young professionals need mentoring and support to grow; experienced professionals need coaching and support to increase performance. Delegation is an excellent instrument to mentor and coach. If you have a young engineer who needs to gain experience and confidence in a given area related to the project, delegate something to him to build up his experience. Delegation with purpose is an excellent and powerful management instrument.

In addition to the permanent staff, your group may have students, post-doctoral fellows, and visiting scientists and engineers. It is important to keep the motivation of your team as high as possible. Promotion of innovation in industrial research is about having a working environment favourable for creative thinking to flourish. Innovative industrial research depends on highly skilled people in an open-minded and supportive environment. Ideas can be promoted by letting staff talk, by disseminating and defending their ideas, or by recognition of innovation.

It is very good to have visitors around. However, just as there are some advantages, there are also some risks you cannot ignore. Risks could come from unethical behaviour on the part of collaborators. This could happen from not recognising the contributions of involved parties, people being recognised more than they deserve, stealing ideas, discussing secret research matters in public, or using the visited organisation's name to their advantage.

If you work in an organisation that aspires to be a source of innovation, you should promote and recognise original ideas while respecting the governance and policies of your organisation. And you should be careful in discussing project details (see Chapter 6 for details).

Working with widely distributed teams is not as big a problem today as it was in earlier times. There are many technologies available to support communication at a distance. However, nothing substitutes for your presence. Reserve sufficient time to visit your team, organise workshops to discuss problems and solutions, get the team talking as much as possible with each other.

Finally, it is worth mentioning that it is very common to have interdisciplinary teams doing research. Science will be more interdisciplinary in the future than it is today. For many generations, scientists have been working in well-defined and established disciplines. This is changing. I understand in the future there will frequently be biologists working with information and communication scientists to solve a problem in agriculture; or astronomers working with geneticists and mineralogists to answer questions formulated by anthropologists; physicists and entomologists solving problems in electronics. Successful scientists need to work across boundaries. Working across different disciplines is and will be even more important for real innovative processes in industrial research. Research should be able to respond to a new way of 'seeing the world', being the driving force to change the way people interact with nature, providing better ways to correct mistakes, all this as well as generating value. Scientists have to be open minded, trained to work in teams, and be ready to capture opportunities to collaborate in order to make a larger impact on society with their research. The crucial aspect in this situation is to give each professional an overall perspective of the project and an appreciation of how important their work is to the project. You should clarify what is expected from each party and why their research is relevant.

RESEARCH STRATEGIES

You cannot be successful in developing applied research if you don't understand business needs and timetables. Once you have a hypothesis to test, you should define a strategy to test it. The strategy is deeply associated with the experimental arrangements you have planned. Everyone in the team should understand the problem, the hypothesis and how you want to test it. Having this three-fold aspect gives everyone in the research team a clear and concise picture of the project. The research phases that require deep intellectual effort are in defining the problem, establishing the hypothesis and the experimental arrangements to test it. This is at the core of research.

The potential for innovation becomes visible in these phases. The best time to judge the possible impact of your research is when the problem, the hypothesis and experimental plan are defined.

RESEARCH OPERATIONS

The research operation is about having the team working together, each person with a clearly defined activity, engaged on the problem and experiments to confirm or refute a hypothesis. Imagine that you are working in the wine industry in Tasmania. You know the problem is, for example, that the wine produced has increased its sugar content above expected levels during fermentation because of the vines' response to warmer summers. Your hypothesis is that if you pick the grapes one week earlier and add a small dose of sulphur to the wine you will decrease the sugar in the wine without affecting the production volume. Research operations are more than just picking grapes one week earlier than expected or adding sulphur; it is about documentation, being present in the field, checking the quality of the sulphur, guaranteeing that other factors are being recorded, checking how warm and rainy the days were this summer, keeping budget and time under control. It is about reporting the progress of the project to your stakeholders. The reputation of your team will depend on well-deployed research operations.

CRITERIA FOR SUCCESS

Research projects must have established success criteria. If your research is about increasing production, you should define how many units you are expecting to produce. If it is about reducing greenhouse gas emissions, you should define how many tons of CO_2e. In the same way, you can define an increase in your research outcomes. The point is: if you don't define success, how can you be sure you're successful? Discuss what success means with your clients and team and agree on a definition.

RESEARCH OUTPUTS AND OUTCOMES

There are many possible research outputs and outcomes – publications, better (or new) products and processes reported and protected by patents, awards, and so on. You should be ready to achieve them all! As a natural product of research activity, it is a good idea to set aside time to work on publications, patents or filling in applications and proposals for awards.

Research outputs and outcomes are different from impact. Impact from research activity is when the adoption of research products produces economic, social or environmental wealth.

Finally, it is worth mentioning that, if the research team can take the lessons learnt from their last project and update a list of these experiences, they will provide future projects with valuable advice. Hold on to what you learn through your projects: try to repeat the good things and avoid the problems. Organisations tend to repeat their mistakes because the same conditions are in place (e.g. governance policies, organisational culture, legacy issues). Lessons learnt by a research team have proved valuable for gaining interesting insights into relevant problems.

LEGAL ASPECTS, BUSINESS DEVELOPMENT AND INTELLECTUAL PROPERTY

You do research in industry because it is expected that your activities will produce some benefit. They might be things like improving working conditions, developing new products, increasing market share with better products or reducing environmental impacts. No matter what your research produces, you will benefit more by paying close attention to the legal aspects related to business development and intellectual property.

You need to understand and seek professional advice on non-disclosure agreements, licences, contract and consultancy agreements, collaborations with other organisations, contracts with suppliers, students and visiting researchers, patents, royalties, copyrights, and spin-offs. Every researcher should be minimally informed about these arrangements, which vary from business to business and from country to country. Legal matters, business development and intellectual property are important aspects of research and cannot be neglected. Innovative research has a much better chance of bearing fruit if proper legal advice is available.

EDUCATION AND OUTREACH ACTIVITIES

In my view researchers should visit schools. Some projects related to space agencies commit their scientists to investing 5% of their time in educational and outreach activities. I have visited more schools than I can remember and talked to hundreds of thousands of students in my professional life. I do not regret a second I have spent with students. The project and the industry will benefit because the public is being positively engaged in and informed of the scientists' activities. Society will benefit because students will come to understand different economic activities in their city. Science will benefit from outreach and education activities because a new generation of scientists will emerge and society will understand more about science in general.

There are many educational and outreach activities you can develop to reach the wider society with your projects. For example, if you work in a large industry you may have hundreds of colleagues' relatives in school with doubts about their future professional career. You can organise to have them working with you or visiting the laboratory to talk to different professionals and learn the pros and cons of their professional activities. Alternatively, you may want to explain the relevance of these projects in the area of influence of your organisation. If you have a robot or a ship, you can run a competition with the students to paint an illustration of the equipment or give it a new name.

Any one of these activities is easy to implement, is cheap and produces great interest for students, which will increase their engagement with science and their studies in general.

The important point is to seriously consider having yourself and your team working in educational and outreach activities. In general, researchers widely neglect their potential contribution to disseminate the value of science to society and the many impacts research can make.

To illustrate important aspects of research management, I'll review some of the ideas of this chapter with an example. Imagine that you are working for a project with the objective of discovering if fresh water is being released from a geological feature located at a depth of 4000 m close to the Antarctic continent. You will need to set out the following as parts of your research project.

- A clear statement of its purpose and scope: The purpose of the research is to discover a suspected continuous fresh water flow under the ocean, and to test the capabilities of sensors and autonomous submarines to work at great depths in a given region of the Southern Ocean.
- The project objective: To discover whether there is fresh water emerging from a geological feature located at a depth of 4000 m, close to the Antarctic continent.
- Hypothesis, experimental strategy and expected results: The hypothesis is that there exists a relevant vein of fresh water

4000 m underwater, close to the Antarctic continent. The strategy is to have salinity sensors on board a small autonomous submarine that will cover an extensive area close to the seabed to measure water parameters. The results are expected to have low temperature, strong currents and substantial variation of conductivity (proxy of salinity) at depths.

- Timeframe and milestones: Because of battery and navigation constraints, each mission with the submarine will last 50 hours, and 10 missions will be undertaken during the southern hemisphere summer.
- Budget and assumptions: It is assumed that infrastructure will be provided for a team of four researchers on board a large vessel with spare parts available, and so on.
- Expected output, outcomes and impact: The discovery could impact on the understanding of water dynamics at depth, and the influence of a large freshwater stream under ice. The impact will extend to the food chain of the entire Antarctic region, leading to better planning of fisheries industries and understanding its implications for climate change. As research output it is expected that many scientific papers will be published (one in the journal *Nature* or in *Science*) in addition to extensive media attention.
- Success criteria: Navigate with your underwater vehicle for a distance of 1000 km collecting useful data during the summer.
- Quality assessment: Sensors are calibrated, the submarine was deployed where expected, nobody got hurt while working on the research project.
- Educational and outreach activities: You run a competition with students in your state to give a name to your autonomous submersible.
- Risk assessment: Spare parts are on board when needed, contingency plans are in place, a reasonable list of potential negative events (such as strong storms) was accounted for and discussed with the team. No data to be disclosed until proper analysis is done and the papers in *Science* and *Nature* are published.

Part 3

Secrecy and openness in industrial research

Chapter 6

When is the work finished?

CONFIDENTIALITY AND OTHER PRECAUTIONS

When an industry invests shareholders' resources in research it is expected that this activity will result in something profitable. Profit here can be understood as increasing market share, reducing costs, improvement of quality, better environmental performance, or even intangible values (ones that are difficult to measure). All investment in research from any kind of organisation can be lost if their researchers are not able to keep quiet about what they are doing. Just like in any family, there are issues in research that can be public and other issues that must be resolved in-house.

If a research group is working on the development of a new product, the researchers have good reason to not speak about it publicly. There are a few reasons for this.

- Competitors will try to launch a similar product earlier, gaining advantage in the marketplace.
- If the product is not launched this might be perceived as incompetence by stakeholders and shareholders.
- Someone else can patent that new product earlier.

If you're working with other organisations (that is stakeholders), the problem may be even more sensitive.

The important thing is to keep everyone informed about the level of secrecy required for research that is being conducted in your department. Establish a policy to clarify why it is relevant and define the expected behaviour of the researchers involved in the work under public embargo.

As a researcher your primary obligation is to protect the investment of your company. If your organisation is involved in research that is not in the public interest, or could harm the interests of shareholders and stakeholders, society or the environment, you have two alternatives in front of you: either you try to convince your organisation to abandon the research or you leave the company. You may consider public disclosure of the information as another alternative. However, before trying this I would recommend you consult your lawyer and your family. You may be bound to non-disclosure agreements and this may harm your career and reputation for ever.

Keeping your research secret requires some precautions such as the ones set out below.

- Clarify with your managers the level of secrecy required for each piece of research you do.
- Certify that everyone working with you understands and adheres to the level of secrecy required for the research.
- Certify that all documents are signed (for example non-disclosure agreement).
- Control copies and back-ups.
- Protect important electronic files with passwords.
- Restrict access to systems.
- Be careful about who visits your institution.
- Don't leave printed copies on your table.
- Erase whiteboards after meetings.
- Destroy documents and handwritten notes after reading.
- Don't even consider talking about the project in public areas.
- Give projects exotic names unrelated to your research (such as Monet, Anaconda, Mammoth, Castle, Saturn, Pluto, Lava).
- Inform the team when secrecy requirements change (such as after a patent has been granted).

If you decide to leave a company for any reason, clarify the level of information regarding the project that can be disclosed in your curriculum vitae. This is because, even though you have left the organisation, you will still need to adhere to non-disclosure agreements. Some organisations require that employees not work for competitors for several years after leaving the organisation.

Patents play an important role in industrial research. A patent is a right granted for any device, substance, method or process which is new, inventive and useful.

A patent is legally enforceable and gives the owner the exclusive right to commercially exploit the invention for the life of the patent. The patent process and type varies from country to country. In Australia, a standard patent gives long-term protection and control over an invention for up to 20 years. On the other hand, an innovation patent is a relatively fast, inexpensive protection option, lasting a maximum of eight years.

Patents give effective protection if you have invented new technology that will lead to a product, composition or process with significant long-term commercial gain. In return, patent applicants must share their know-how by providing a full description of how their invention works. This information is made public and can provide the basis for further research by others.

Of course there are things you cannot patent such as mathematical models, plans, schemes or other purely mental processes, including artistic creations.

SCIENTIFIC DOCUMENTATION AND REPORTS

Internal reports

When a piece of research comes to an end it might be worth documenting the results in a report. Reports are no different from any other written document – you should have a clear idea of the purpose of the document, who is going to read it and why they should care about it, how long it should be and its format, and you should pay attention to any circulation restrictions.

Research reports are usually structured around a number of elements.

The cover should have an attractive format, showing the title, name of your organisation and department, authors, and date. You may want to inform the reader upfront if the document is secret (such as by using a watermark). The back cover may contain disclosure information and acknowledgements.

The report begins with an executive summary. This should contain no more than two pages. Concentrate on results and actions following the obtained results. If executives in your organisation want to know more, they can look further into the report where all the technical details are shown.

The document follows on with an introduction to the motive for the research, the questions to be answered, the problem that was studied and a summary of the bibliographical survey. Then the experimental results are given, followed by analysis and discussion. Conclusions and recommended actions should follow (this part will be more extensive than the executive summary).

Finally, you can finish the report listing the literature that was consulted and an appendix (if necessary). In the appendix the researcher should put any supporting material like additional graphs, mathematical demonstrations, detailed equipment description, and forms used in the research. The appendix may also contain a CD with raw data, photos, video or sound recorded during the research.

I recommend that you read a few of your organisation's reports before you start writing your own. If no model or template is available, take this opportunity to create a new standard of report writing in your department!

I know some organisations value technical documentation very highly and write reports on even very small pieces of work or meetings. Some of these reports contain only two or three pages; they are filed away but are easily accessible through keywords.

Another reason for writing a technical report could be if there is a new operational procedure – this could be a description of a given operation.

Almost all research organisations have well-structured documentation centres and you should both use and contribute to them. Finally, I must emphasise that time spent in writing (really good) documents is a terrific investment.

I have a final recommendation: respect copyright. I am still shocked when I see some large companies photocopying books or scanning them entirely, and distributing them internally as an annex to their internal reports.

Public reports

Public reports are usually more general than internal research reports. The format and content, as expected, depends on the audience and on what can be published. It could be a folder containing information about the capability of your research centre, references to publications, list of clients or partners, and images. It could be a document meant to be downloaded from a website and printed. No matter the model or venue, you should perfect creating public documents. Being perfect means being precise in the information you disclose, avoiding typos, having a clean and attractive design and being in agreement with the organisation's policy. Copyedit documents carefully.

The image of your organisation strongly depends on what you publish.

If the information goes to the website of your organisation, it is a good idea to check its content. Verify that each link works well and that you are respecting copyright in linking different sections. If you want to present a link to a paper you have published, provide its DOI (document object identifier), not a file. DOIs provide efficient links to the abstract of your papers in the publisher's website and they follow the copyright policy of each publisher.

Internet tools offer a great deal of resources you can explore for a reasonable price. You can prepare animations, podcasts, documents, and links to additional information.

The important aspect while planning a communication plan for your research is to consider your target audience, their interests, and what you want to achieve with the communication. Content, format, and media all come later.

RESEARCHERS AND COMMUNICATION

It is easy to find technological research that has some scientific merit. Developments and solutions achieved by industry interest the scientific community because, among other things, they represent the best examples of application of their method. This fact alone is enough to justify more investment in fundamental research. Another compelling reason for academic researchers is that they like to talk about possible applications of their area of study. Researchers are proud of their research and passionate defenders of their work. They will even discuss applications that are created by other researchers! Therefore, a new industrial application that comes out of their research area is absolutely relevant for discussion in academic circles.

The quality of research can be assessed by its innovation, hypotheses tested and the results obtained (such as breadth, depth and precision). Going beyond its quality is how successful researchers are in reporting their work. As well as publishing in the research area, all the above aspects will determine whether you can target a high quality journal or conference to publish your research. If your industrial research is of general interest and has the attributes mentioned above, you may consider publishing your paper in high impact interdisciplinary journals like *Science*, *Nature* or *PNAS*.

If you don't know exactly where to publish, look at some of the journals in your area of interest. Read them. Check their impact factor. Scientists who ask for grants or apply for jobs in research centres will have their science merit judged primarily by the number of papers and the impact factor of the journals that published their research. There's no short cut or excuses: good researchers are expected to publish in good journals.

Each journal has an impact factor that can be calculated from the number of times their papers are reported in the literature. Set yourself a challenge to publish the results of your research in the best journal you can.

The editorial process is reasonably similar for all journals. In general, the editor will verify that the content of your paper meets their editorial policy. If it does not, the editor will let you know their decision quickly; and may suggest you submit the paper to other more appropriate journals. Once an editor identifies the work as acceptable for their journal, they will send the manuscript to a number of referees. These peer-reviewers will read the manuscript carefully. They will assess your paper against the following criteria:

- relevance and novelty associated with the problem investigated
- methodology and experimental procedure

- extent and quality of the cited literature
- quality of the text (orthography, clarity, logic) and other aspects that might result in a limitation of the work or of the research value.

Referees are usually quite harsh in their reports. Their identity is usually kept secret.

You will receive some form of communication from the editor showing the referees' comments on your work. It is up to the editor to decide, based on their comments, if your work can be published in its current form, if a revision is needed and how extensive this revision should be. Of course the editor can decline to publish. None of the referees' comments will come without a justification and often there are similar comments from each one. However, if a difference of opinion occurs, the editor can decide to publish or call upon another referee to solve the dispute.

The important aspects in this evaluation process are:

- science has as its core values: the openness of results and free discussion of methods, hypothesis, experimental results, and so on.
- it is time-consuming. Few journals offer a quick response (varying from days to weeks).
- the refereeing process is not perfect. In some cases a referee's report may give the impression they don't understand the work, but even then it is not for the authors and the editors to argue.
- in most cases, it improves the quality of the work.

The fact that the publication process is time-consuming can be frustrating for young scientists, especially those who would like to have their results published before they have to defend their PhD thesis. However, in my experience I have never found a single referee comment to be useless. Take responsibility for writing your paper as clearly as possible. If a referee cannot understand a piece of your work, it is very likely other scientists will have the same problem.

Let me tell you a secret: while doing research, you can come up against a situation or a problem to which there is no evident, firm or good solution. In this case, researchers can do one of three things:

- they do not publish the work

- they present a paper at a conference expecting good feedback from participants who will express ways to improve it
- they submit a premature paper to a good journal.

If you decide to try for publishing in a journal, explain your situation to the editor. Bear in mind that you are not looking for publication, but for a formal opinion of a peer-review panel. Referees of good journals are usually very rigorous in their report. They can frequently be sarcastic, or even appear ill-tempered in their remarks. In this case, it is important that you recognise their contribution: either as a co-author – if they deserve and accept this, or in the acknowledgements or reference section.

A final piece of advice: referees' criticisms are about your work, not about *you*. Follow their recommendations and better work will ensue, or at the very least a different hypothesis will be tested and compared.

Papers you and your organisation publish are truly like business cards. When approached by other researchers or clients, you can offer them pre-prints of your papers. You and your department will be contacted by other professionals requesting information about your work (such as a copy of a manuscript, references or even samples and raw data) and you will be invited to present a seminar, talk in conferences or submit review papers to specific journals, and editors will contact you requesting your support as referee of other researchers' papers.

In fact, the entire process involving scientific publications is intense, but rewarding. I have learnt something from each paper I have published or revised.

And as members of the scientific community we have no excuse but to support editors with the peer-review process (that is with quality and timely feedback).

I learnt the value of collaboration in research early on. Of course you can develop your research working by yourself and publish as a single author. However, exchanging experiences, hosting graduate students, visiting other labs, writing papers together – all contribute to the quality of your work. I prefer working with a collaborator to working alone. Your collaborator could be a colleague from another department, a client, or a scientist from another industry or public research organisation.

Another important point: scientists are well known for their poor presentations! The quality of your figures, carefully considered content and professional presentation is as important as the technical details of your research. It is quite possible that a technical work of high value but poorly presented will be taken less seriously.

There is a huge amount of literature and training available to support you with presentation techniques and to build your skills in preparing presentations.[1] The important thing to remember is, whether it is in a conference or internally in your organisation, quality of presentation is as important as the content – poorly organised presentations are associated with poor quality work. Bear in mind that you will be judged not only as yourself, but also as a reflection of the organisation you represent.

The internet

If someone asks you to find information on a given subject, I am sure the first thing you do is search the internet. You can even do this on your mobile phone. So naturally you should consider the internet a good place to disseminate your work.

In this medium, there is no need for you to present technical details of your work. However, you should describe, taking all due care, what your department is doing, the problems that have been solved, tools and methods you have applied, a list of publications, conferences you and your colleagues have attended, collaborations, awards and other recognition received, key contacts, positions available, and so on. The costs associated with setting up websites are very small compared to other media alternatives.

If your department has no website available, propose creating one. Access the website of similar organisations and compare the content (for example what is shown, update frequency, breadth and depth of information available, and links).

The internet has changed the way you access data and information. Provenance and quality of whatever you access on the internet are important issues.

Publication in journals is a key step in the scientific method. It includes a peer-review and careful consideration about the purpose of publishing results of research. Disseminating knowledge is integral to a researcher.

Part 4

The researcher's responsibilities

Chapter 7
Ethical issues

INDIVIDUAL AND INSTITUTIONAL REPUTATION

Society constantly challenges individual and institutional reputations. So it should come as no surprise that society expects highly ethical behaviour of researchers and those organisations that conduct research. People involved with research should be exemplars of ethical behaviour!

When a researcher fails to adhere to ethical standards or policies established either by government or their organisation, the reputation of the individual and the institution is at risk. Researchers must understand that poor practice and misconduct in science potentially damage the organisation and the entire research body.

As a researcher, you can help build a good reputation if you:

- maintain high standards of ethical behaviour
- spend public resources responsibly (such as grants and scholarships)
- carry out a high quality and honest analysis of your data
- give due recognition to the contribution of others
- disseminate high ethical standards to a new generation of researchers
- act when you suspect or know a colleague has breached ethical behaviour.

On the other hand, poor quality research, deception, and dubious conduct when conflicts of interest occur destroy an organisation's reputation.

It is an interesting question – why do researchers fail to adhere to ethical standards? I argue that transgressions in ethical standards occur mainly because of lack of awareness (i.e. there is a lack of education and frank discussion about ethics in science), pressure for results (i.e. self-promotion, being promoted to higher positions or being more competitive in obtaining grants), and individual character.

It is surprising that some journal editors do not respond to reports of plagiarism. This may be because they do not want to deal with all the stress associated with these kinds of investigation or because of the bad publicity that would be associated with the (true) impression of their poor review process.

In the following sections I will discuss some relevant ethical issues in science. I will finish the chapter by analysing how scientific knowledge and technical development have pushed ethical practices to new limits. I will also touch on issues where scientists are tempted to discuss politics.

CONFLICT OF INTEREST

A conflict of interest is a situation where a professional or an institution has competing personal, professional or institutional interests. In such circumstances your professional judgement can be compromised. Policies and procedures are usually designed by research organisations and scientific journals to protect the integrity of conducted research, the interest of stakeholders and, further, the reputation of everyone working in science.

Conflicts of interest can put your reputation at risk. Because your professional decision can be considered influenced by other interests, you can be accused of ethical transgression. The best attitude is to clear the conflict of interest as soon as possible.

There will be situations where you will need to decline revising an article or a grant application, taking part in a hiring process, or accepting a research position. An experienced researcher should have the ability to judge when a conflict of interest occurs, clarify the issue with those involved and remove themselves from the situation. The consequence of a conflict of interest is not worth your reputation.

The best way to handle conflicts of interest is to avoid them entirely. For example, imagine that you have to buy equipment to conduct an experiment and a few companies respond to your invitation to tender for their products. Among them there is one where a relative of yours works. In this case you should either remove the company from the list of potential suppliers or ask someone to take your place on the purchasing committee. It is important for all involved that you disclose the reason for your refusal to sustain a situation where a conflict of interest occurs. It is vital that you understand there is no middle ground solution for a conflict of interest. Being accused of professional misjudgement is a nightmare.

MISCONDUCT IN RESEARCH WORK

Misconduct in its diverse forms (e.g. fabrication, falsification, plagiarism, self-plagiarism – 'déjà vu') and the rule of intention will be discussed in this section.

Fabrication can be understood as the making up of data or results (e.g. creating data without a real measurement process). Changing results to favour a hypothesis or deleting experimental data that does not fit to a model without minimum consideration are considered falsifications.

Anyone can breach ethical standards by reporting in their CVs publications that do not exist, activities they have not been involved in, positions that they did not hold. Citing references as 'submitted' without submitting the paper, or as 'accepted' without a formal acceptance letter, or 'inflating' citation records, especially in grant applications, are serious ethical transgressions.

Misleading authorship credits is a serious and very common breach of ethical standards. Typical misdemeanours include listing authors without their permission, attributing work to others who have not in fact contributed to the research, and the lack of appropriate acknowledgement of work primarily produced by a research student/trainee or associate.

Misconduct of research can lead to substantial loss of time for the research community, forfeited recognition of others, and a devastating overall feeling of personal betrayal among involved staff.

I encourage you to visit the publication website of some academies of scientists and read their recommendations regarding ethics. For example, the Australian Academy of Sciences classifies ethical misconduct as misrepresentation and misappropriation.

Misrepresentation occurs when a researcher deceives or has a reckless disregard for the truth. Misappropriation occurs when a researcher plagiarises (e.g. presents the documented words or ideas of another as his or her own, without the appropriate attribution for the medium of presentation); makes use of any information in breach of any duty of confidentiality (e.g. by having access to original work when doing a review of a manuscript or of a grant application); or intentionally omits reference to the relevant published work of others for the purpose of inferring personal discovery of new information. You should always take great care to clarify what your contribution is and what part of the paper is the work of others.

Self-plagiarism or duplication is another (common) problem in research. There are researchers who publish their research twice or even more often. Sometimes they change the order in which the authors appear as well as the title of publication. Fortunately there are several instruments available to identify duplication. Some types of software are now available for the detection of plagiarism – some of the most popular being iParadigm's Ithenticate and Cross-Check, EVE2, OrCheck, Copy-Check, eTBLAST, and Déjà vu. Another easy way would be to copy parts of the text and search at Altavista or Google. Plagiarism and duplication clearly violate copyright and should be managed with rigour.

Some scientific societies are considering publishing cases of duplication, publicising the name of those identified in transgression as a mechanism to punish bad ethical behaviour and create awareness of the desired responsible conduct of science. In other cases, editors have been reluctant to investigate transgression in science because they want to avoid the stress involved in a scrutinising process, and others are afraid of revealing their poor editorial processes.[1]

A researcher should not take, deliberately conceal or materially damage any research-related property that belongs to another. Included here are experimental apparatus, reagents, biological materials, writings, printouts, data, hardware, software, or any other substance or device used or produced in the conduct of research.

If you come across any case of ethical transgression you should act at once. The researcher involved may then be able to use this opportunity to correct their conduct. Sooner or later, these cases will come to light and will damage the reputation of the scientific society, the organisation and other stakeholders.

ASSIGNING CREDIT FOR WORK

Fraud may be the worst sin in science, but mistakes (intentional or not) in allocating credit and responsibility for work are also serious. In the standard scientific paper, credit is explicitly acknowledged in three places: the authorship, the citation of someone's work and in the acknowledgements. This section will discuss how the reader can acknowledge work done by others and help them understand the importance of making the researcher's and a third party's contribution to the work quite clear.

Some scientific journals are now adopting interesting ways to assign credit within co-authorship. *Nature* requests authors to declare they have contributed equally to the work. *Proceedings of the National Academy of Sciences* (*PNAS*) request authors to indicate their specific contributions to the published work. This information will be published as a footnote to the paper. 'Contribution' is listed as designing research, performing research, contributing new reagents or analytic tools, data analysis, or writing the paper. Each author must explain what they have contributed. Of course one author may have contributed in more than one way, while others may have contributed to the same aspect of the work.

A conflict of interest may also arise if you are asked to review a paper that presents advances in the same problem you are working on. In other cases, a researcher can be extremely generous in sharing original research with other scientists. They might be doing this because they want to share ideas, explain the results they obtained to help other researchers in their work, or for any other reason. No matter why the researcher decided to share their original work, you should keep it between you and your colleague. However, if you are going to publish a result that may contain part of their ideas, an experimental arrangement or even one of their results, you should consult them. You are not allowed to publish your work without their consent, even if you cite them in a private communication. I would strongly recommend you seek their permission or advice regarding your publication and establish how you should recognise their contribution. Again, you have three options to offer them: co-authorship, or a mention in the references or acknowledgements. If you publish work containing their ideas without consulting them, even if the possibility has been discussed between you in your own office, over dinner or on a flight, your action will cause them to believe that you have stolen their idea and your colleague could call you unethical. In this situation, your colleague would be right.

NEW ETHICAL PARADIGMS

Science is able to solve serious problems and I hope it will gain more importance in society. For a scientist of the 21st century, I can imagine wonders like genomic engineering guiding customised medical treatments; or machines integrated into the human body, monitoring and substituting for large organs; nanotechnology copying viruses and creating new ways to protect us against diseases, discovering and exploring planets, better and cheaper communication processes, and minimising the human footprint in the environment.

The tremendous progress of science will push ethics to limits unknown to us today. For example, recent publications show evidence that there are some psychological disorders associated with specific human genes.[2] Soon there will be commercially available tests, at affordable prices, to check if a patient has the gene indicating a predisposition for developing a psychological disorder. What would you do if you knew you had a genetic predisposition to develop, for example, bipolar disorder? Would you have children? In the future, will companies ask for genetic analysis before hiring? Will insurance companies increase your fees if they have access to your genetic data? Will this be ethical? This is a simple example of how scientists should be aware of the consequences of their research to society.

Researchers should be proactive, capable of predicting ethical issues and providing leadership to those adopting their inventions by putting in place ethical solutions. Scientists should be able to address new ethical paradigms before they put out processes or products for adoption.

POLITICS AND SCIENCE

In my understanding scientists should be free to conduct their research, promote their findings and participate in discussions related to issues on their research. Scientists must have the right to speak and present their research. However, the best venue for this discussion is the scientific conference, not radio, newspaper or television.

Researchers should be extremely careful when commenting on matters of public interest. When the subject is a matter of contention and sharp debate they should provide proper information (not opinion or beliefs!) and avoid commenting on public policy. This is particularly important for researchers working for public organisations.

Researchers have the right to speak freely. However, they are not usually familiar with the world of politics and should be very careful in expressing their 'opinions' on public policy, recognising that the government remains responsible for the articulation, formulation and implementation of government policy. As a researcher working for either a public or private research organisation, you may be invited to contribute to the formulation of public policies. It is good to see that researchers can contribute to public policy with their knowledge and they should be encouraged to debate on scientific and other research issues of public interest in areas of their expertise. They should be fully supported in open communication, the debate of scientific results, and the dissemination of information.

Politics, however, has its own agenda and methods. While researchers are trained to see the world in a very objective way, politicians have different needs. The debate of ideas in a scientific forum is very different from that in the political arena. Bear in mind that whatever you say on TV will be taken not as your own views, but those of the organisation you work for. Therefore, you must be very careful to ensure you have been given permission to speak and, before speaking out, seek appropriate support from your institution. Again, limit your discussion to the realms of research.

Chapter 8

Values in research

NOT ONLY KNOWLEDGE ...

Research is carried out to improve people's understanding of nature and to benefit society through the application of new knowledge. But apart from this primary benefit, there are more rewards to be gained from research. For example, an organisation that is known to have an innovative culture in place has a good image in society, has overall more motivated staff, and is sustainable.

Private organisations whose brand is associated with research, development and innovation are likely to be successful. There are subtle but important differences between efficient organisations and innovative organisations. Efficient organisations and their teams are competent in achieving their targets all the time. They may have a strong culture associated with total quality programs. (Skilled managers may provide strong systematic procedures that could even block innovation. They are aiming at efficiency, not innovation.) On the other hand, organisations open to and engaged with industrial research have innovative products and services; they achieve targets nobody has seen before and gain market share or create new markets.

Efficiency in organisations is important (i.e. routine management: 'hitting the target'). But an efficient organisation without innovative processes (i.e. innovation: 'hitting the target nobody has seen before') is on the path to obsolescence. Another benefit an organisation gets from research is that it can increase their sustainability.

One of the objectives of industrial research is enabling private organisations to be sustainable. Sustainability is based on actions that have economic, environmental and social values. While carrying out industrial research, the social impact of the work is taken into account. The bottom line here is that industrial research with economic, environmental and social impact is innovative, and has more value. The benefits of research activities are far more than the knowledge alone.

In fact the benefits usually go beyond what was expected. For example, new materials developed for space exploration are now available in our daily life. Cordless power tools and appliances you use today were developed for NASA's Apollo program, to support astronauts drilling down beneath the moon's soil to collect core samples. Consequently, the drill needed to be small, light and battery-powered. Smoke detectors that were developed for Skylab missions save many lives every day, and even the material used in surf boards was developed for space applications.

However, it must be said that even if research outcomes have provided us with better products, materials, dwellings, medical treatment and understanding of nature, these outcomes have not benefited everyone. The benefits of all investments in research have not reached communities and countries equally. While protection of intellectual property is imperative in business, it is important to guarantee the impact of research on society as a whole, and that reality is still far away.

RESEARCH IN PRACTICE

Industrial research has value when it is applied, and can improve the sustainability of the business. Therefore, industrial research has value only when it creates impact. Applying the discovery is more important than the discovery itself. So researchers should constantly be applying their research and creating a positive impact with their findings.

I have seen many inexperienced researchers, old and young, neglecting or undervaluing the impact of research. How their research work can create impact should be clear from the beginning of the research project. The impact could be cost reduction, quality improvement, a new material, better understanding of processes to reduce pollution, a new medical treatment ... you name it! The point is: the desired impact should be clear right from the beginning.

An industrial researcher who produces only reports, literature reviews or other documents, and no good journal papers, patents, improvement of processes or real impact from their research output and outcomes should start to consider looking for another job.

The value of research is when it is applied and it brings benefits. To put industrial research into practice is not easy and requires researchers to understand operational constraints of industrial plants, performance of equipment and processes. For example, if a researcher is working on developing a new mineral agglomerate for sintering processes, they must understand the capacity of the industrial plants, the limits of the furnace, or even the impact of the new raw material on the emission rates of gases and particulate matter, and so on.

No industrial research should have as their objective simply science *per se* or knowledge that does not have an application. Science and acquisition of knowledge is a way you can create impact from your research activity.

I'll give you an example. Suppose a senior researcher in an oil and gas industry has been asked to support the environmental department of their company with their research. They developed a project to investigate the concentration of airborne particles in an industrialised urban area where the company operates. They discovered that the concentration of pollutants increases on average 40% from the mean value during dry afternoons in summer. They found the result interesting and published their results at an international conference (frequently considered 'scientific tourism') and in a journal. But they did not use these results to attempt to change the industrial activity or traffic management in their area when the particulate matter concentration is known by their study to be critical. They simply wrote a short report for their company and moved to another topic. What is wrong here?

What is wrong is that the impact of their research is questionable. Actually, there is probably no impact from their research: at least for their employer and the city they live in! There is almost no benefit to society, the economy or the environment. The researcher's work has had no impact: this is a typical consequence of falling into the temptation of researching in the dead-end of knowledge *per se*. The researcher is, as some researchers are, isolated from reality.

You should always ask yourself a few questions before you start a research project:

- 'What is the objective of this research?'
- 'Why it is relevant?'
- 'Who will benefit from the research?'
- 'How can I apply the knowledge I gain from my research?'

There is no real innovation in industrial research without impact. You must consider the impact of every bit of research you do – this is the way to achieve professional success as a researcher. Your publication record is not a short-cut to success: it is the long road ...

Good researchers should also be good project managers. The design and implementation of innovative projects require that you have a minimum of knowledge about project management. An experienced researcher may have a group of young researchers working with him: students, post-docs, young scientists and engineers. You have to manage people and research project outcomes. Don't forget that great research projects will require good management practices. In reality, most researchers lack management skills, but those who possess them are usually successful.

To sum up – it is important to remember that research activity offers many opportunities. Working in research can be rewarding, but you will be asked to adhere to rigid ethical and moral values. Your written reports or public presentations or interviews to the media should be honest; you should be clear on what your contribution is and what the contribution from others is. You should declare the limits of your analyses, conclusions, and uncertainties involved, the constraints of your hypothesis and what is still unknown and yet to be done.

The claims in favour of your hypothesis should be impartial. There are two important reasons to be absolutely honest when doing research.

- Researchers don't quest for truth with lies or half-truths.
- Misconduct in science affects the entire scientific community and primarily the involved organisations, and everybody working there.

It is not easy, but you can be well rewarded for your efforts.

Part 5

Innovation in industrial research

Chapter 9

Is there a need for innovation in industrial research?

Everything can be improved. Ways of thinking, processes, products and services can be better tomorrow than they are today. In this chapter I will discuss problems in current industrial research practices and arguments for the need for innovation. I will identify some limitations in today's industrial research and prepare to discuss solutions to some roadblocks that will be covered in detail in Chapter 10.

WHAT DO ALL INDUSTRIES NEED?

Any type of industry needs to be sustainable. Being sustainable is consistent with promoting social, economic and environmental wealth from activities. All industries need their clients and those who use their products, a perception from society that they have value, and employee satisfaction. The industry needs new markets, better and competitive products, and cost-effective processes. Industries today need to improve the environmental performance resulting from their operations, and those that cause less detrimental impact on the environment will be penalised less. Industries should aim for cleaner processes (e.g. generating fewer and non-hazardous by-products) and be efficient when using natural resources (e.g. water, raw materials, including paper) and energy.

Industry needs ethics in its business, transparency with stakeholders and shareholders, and respect for partners, clients and contractors. Last but not least, they need to look and work for the future. Today this can be done by being sustainable, ethical, and able to influence or ready to encourage changes in the legislative scenario. Through foreseeing other risks, the industry needs to grow and eventually be ready to change its core business. A mining company today might be an oil and gas company tomorrow and later an energy company.

There is no doubt that it is a herculean task, but all these needs are absolutely consistent with good management practices, good people and innovation. Innovation can support the genuine industrial needs of today and tomorrow. In the next section I will discuss how innovation can support industry to meet its needs in the areas that constitute sustainability.

INNOVATION DRIVING SUSTAINABILITY

Sustainability is understood in industry as the desired readiness of organisations to survive. To be sustainable, companies rely on improving their performance in three key areas: environmental, social and economic. An organisation that is not profitable, does not contribute to the society in which it operates or that has significant negative impact on the environment will not survive. For the scope of this book, I will explore the industrial need for sustainability and illustrate how innovation can play an important role.

Environmental

There is no doubt that industries are expected to reduce their impact on the environment. This means, for example, that an actual set of environmental indicators should be continuously improved to reach limits never seen before, and then taken further with new productive processes. Coal was once considered a fantastic source of energy because of its relative abundance, low cost of exploration, easy handling and transport, straightforward use, and so on. Nevertheless, it contains concentrations of sulphur, arsenic and other substances that have raised concerns about its use. Acid rain became a growing problem where the dispersion conditions in the atmosphere were not favourable and contamination of soils and of alluvial sediments increased to toxic levels. More recently, consequences related to the increase of greenhouse gas concentrations in the atmosphere have changed the way the world defines its energy matrix. Coal, once extremely attractive, is seen as a less attractive source of energy. This issue is pushing industries to be more innovative. Some are considering developing better desulphurisation facilities, storing carbon dioxide from combustion back in the ground or burning methane in coal mines (methane has a much higher potential to increase atmospheric temperature than carbon dioxide). None of these technologies would have been proposed or seen to be more efficient without innovation. This is one example of how innovation is important to the sustainability of industry. You may consider visiting the BAT documents (that is best available technique on environmental monitoring and control)[1] from the European Community to see how important advances in technology are and the implications they have on the way we produce goods. Dioxins, inhalable particles, asbestos, indoor air quality, remediation of soils, gases causing the depletion of the ozone layer – the list of problems to be addressed by industry continues. There is no way to address all these, and future issues, without innovation. As I said before, society expects the production of goods and services to be more environmentally friendly. Innovation will play an important role in achieving new environmental performance while preserving the economical attractiveness of future businesses.

Social

No industry is expected to remain viable without having genuine concern and respect for social interests in its area of operations and taking action towards preserving them.

Poverty and inequity in many countries, despite strong economic development, are critical challenges for sustainable development and regional stability. Poverty has been seen in many ways. Many people suffer from malnutrition and have no access to health services, sanitation, clean drinking water, shelter, electricity, modern communication channels or transportation. These are the 'basic' needs of any individual, and with the increasing spread of urban centres the world's populations are experiencing a growing demand for energy and natural resources at unprecedented levels.

The benefits of the information and communication technologies revolution are not shared by everyone. High data volume transmitted in shorter times is creating a revolution in public services, sensor networks, and in the communication between people and organisations. This revolution is changing the way families interact and social dynamics.

This extremely dynamic society is eagerly demanding innovation in the way they communicate, are informed, interact, form groups, follow trends, express themselves and see the world. All this means that all industries must be able to capture, adopt and create innovative ways to develop their businesses.

An industry that is not adapting accordingly to address social demands in their area of influence is sentenced to shut down its operations.

Economic

There are many reasons for economic pressures to drive the need for innovation. For example, if your company wants to be more competitive (such as by having more cost-efficient production methods), explore new markets and seek a larger market-share with better production, it needs to invest in research.

Innovation has been a driver for industrial transformation and continuous success. While incremental innovation creates a competitive advantage, radical innovation creates a completely new market. For example, the first television used rotating discs (mechanical television) to generate images. Then the technology using the principle of the cathode ray tube, invented in the late 19th century, was adapted to television. In fact, television is a confluence of many technological discoveries (for example the cathode ray tube, understanding of radiation, electronics – the first ones used valves). With the invention of the transistor, a masterpiece of technology that changed the world, and with 'incremental' improvements, televisions became smaller, more stable and of better quality. Colour television came later from additional incremental innovation and the screen became flatter, producing less distorted images. The principle is the same: cathode ray tube, improved by successive incremental innovations. The plasma display monitor was developed during the 1960s and followed three decades later by the liquid crystal display (LCD) creating a 'radical' change in television quality. For many decades television was driven by hundred-year-old technology. This 'disruptive' innovation created new products and markets. So, radical innovation creates a new concept and sometimes a new product (and therefore markets), while incremental innovation improves existing concepts and products.

I don't have space here to go into the classical model of early adopters, innovations, maturation and decline of a technology with the advent of a new disruptive solution. The bibliography at the end of the book can offer you a better analysis than I can give you. However, I would stress that industry should pursue innovation if they do not want to be followers, to the extent that they miss the innovation wave and disappear. A sustainable industry is an innovative industry.

THE FUTURE

The ethics involving research will be pushed to new limits because social motivation based on ideas of right and wrong, moral values and rules are changing.

I have already discussed industrial needs as drivers for research, but innovation in industrial research is about doing research in a different way. There are three questions to explore on this topic, which goes beyond what industries do to be innovative and which is discussed in Chapter 10.

- Is industrial research of a good enough standard?
- Is there a need for innovation in industrial research?
- How can innovation be promoted in industrial research?

My understanding is that industrial research can be better. Researchers in industry have been extremely limited by paradigms and the world as it is presented. I will give you an example. For many years, teachers taught classes using handwritten slides. Projectors were a very common piece of equipment, found in every classroom. In the early 1990s a new product substituted for the slides by using a transparent projection of a computer screen placed on top of the projectors. After only a few years a new product came to replace the slide projector – a beamer (that is a computer projector). This equipment is widely used today. The industry was not able to move directly from slides to the computer. There was a paradigm in place: everyone had to use projectors.

Industrial research in general has demonstrated its limitations in introducing radical technologies. These limits might even be self-imposed by researchers. Certainly creative thinking, the courage to propose and explore the new, the quest for hidden truth and a sense of highly developed observation do not appear very often among industrial researchers. The increasing bureaucracy in the modern research world is also killing creativity. Education is not oriented towards the exploration of creative thinking, expression in different art forms and the debating of ideas.

An industry that aims at being innovative should let its scientists explore, debate, propose 'crazy' ideas and be challenged. The majority of researchers cannot see beyond their reality (e.g. the projector used in classrooms) and, therefore, their ideas will be limited to what they see today. Their research will be as limited as their minds.

An innovative solution might come from an original and multidisciplinary approach to a problem. An active and productive research community is improved if it has time to think, a good working environment, the freedom to talk and discuss, and communication across disciplines.

Take for example the way people fly today and compare it to 40 years ago. This business has changed very little, even though the aircraft industry has incorporated many innovations during the last 40 years. Navigation using GPS, efficiency of aerodynamics, longer autonomy, communication on board and safety have been brought to new levels, to mention just a few of the changes. But people are still flying pretty much the same way they flew decades ago. The same is true for cars in that only now are there new propulsion engines available. Within this scenario our generation is suffering from the limitation of existing infrastructure as products and services become popular. I have mentioned two examples here related to transport. There are many others in mining, steel production, energy, manufacturing, and so on. These sectors are likely to suffer pressure to be more sustainable because they have been less innovative. The research departments of industries in these sectors are very limited in scope and are unlikely to produce a breakthrough process. Agglomeration and reduction of ore use very old processes that have been improved with incremental innovation. 'The new way of moving across long distances' is still to happen within these businesses. The way research is conducted needs to be changed if researchers want to accelerate the creation and adoption of new products and services.

So, to answer the three questions: I would argue that industrial research has not been good enough and therefore there is a need for innovation.

Given this problem, what can be done to promote innovation in industrial research? Well, how about reading the next chapter?

Chapter 10

Ways to promote innovation in industrial research

I started this book by describing what motivates scientists to do research and the benefits for industry in doing research. Then I described tools for doing research – the scientific method, how it can be implemented using quality tools and how you should be careful in analysing your data. Then I discussed research management, the communication of research outcomes, and how scientists should responsibly conduct research. Finally I discussed some values in research and the need for innovation in industrial research.

In this chapter I will describe what's to be done in order to be innovative in industrial research.

As mentioned in several places in this book, research is conducted to address a problem. In industry these problems are usually related to ways of improving financial, social and environmental performance. Research conducted in the public sector ideally should address national challenges beyond a single industry's interest, or promote innovation to companies that are unable to conduct their own research. More importantly, because it is the duty of government to solve distortions of the market, it should use research primarily to improve areas that are neglected by private initiative.

There is a genuine need for innovation in industry. But how can the way industrial research is conducted be improved?

To improve and innovate in industrial research I believe that the way research is managed has to change. I will list and explain 10 aspects I believe are valuable and potentially capable of bringing about this change.

1 *Researchers should be paid according to their performance*

I am in favour of paying a fixed salary to researchers with bonuses based on their performance. This will discipline the researcher who does not achieve and will give incentive to the high performing ones. Productive researchers feel that they are not given due recognition if they earn the same salaries as those who perform poorly or if their salaries are only raised by a few percentage points a year. If a researcher changed the way the industry runs its business, produced a competitive advantage or increased sustainability of the business, they should be paid accordingly.

2 *Change the performance indicators of researchers*

Researchers should be asked to provide the industry with new ideas. They should have as part of their performance indicators the production and successful implementation of original, valuable ideas. I argue that research outputs and outcomes are not sufficient to capture the degree of innovation. I say this because I see poor quality papers and useless patents being published every day. And worse, plagiarism and duplication are inflating many researchers' lists of publications. Another problem is the increasing practice of self-citation. Researchers are paid to change and explain things. Publications and doing research itself are not ways to create impact. A researcher who is not able to provide useful results should be dismissed.

3 *Provide researchers with training in business principles*

Universities have neglected to impart business education to graduates. This has created researchers who have poor understanding of the basics in commercialisation, project finance, and so on.

4 *Researchers should be asked to change the industry's business*

Executives should not ask for less. Researchers should be challenged at the business level. Unfortunately researchers will always be asked to do some kind of work that will not produce innovation. Researchers should work 'investing' their time on something that has the potential to create impact. Ultimately, researchers are responsible for determining what they are capable of doing and pushing themselves as far as they can.

5 Peer-review proposed problems to be studied and their associated hypotheses; don't give freedom to researchers without a reasonable cause

Freedom is not only needed, it is necessary. But freedom should not be granted without a peer-review process. Limiting the research to 'core business' or disciplines is blocking creative thinking and stopping new ideas flourishing. Many researchers complain they are limited in their opportunities to work because the organisation does not support them to explore their own ideas. In these cases, the organisation and the researcher are wrong. An interesting problem to be investigated or a hypothesis to be tested should be peer-reviewed. This assessment should have clear criteria (which depends on the industry's interests) and be defended to the executives of the organisation. A researcher asking for freedom to do their research without any judgement of value is against the scientific method itself. Providing experienced researchers with problems (and the means) to solve them is limiting; however, there is also a limiting factor to innovation.

6 Take risks

Most researchers are too conservative. They prefer working with something that they can definitely achieve. This is fine once in a while, but not always. Approach this as development, not as research. Many research centres are doing development only, something that engineers can certainly do better then scientists! Researchers should take risks and try out their more interesting ideas. For example, would you ever try to land a robotic vehicle on another planet using airbags? How many risks did researchers working at the Jet Propulsion Laboratory in Pasadena, California, take to successfully land three rovers on Mars using airbags (Sojourner, Spirit and Opportunity)? You will not be successful in research if you do not take risks.

7 Researchers should experience working under stress or difficult conditions

Researchers should always receive less than they ask for. They should frequently be taken out of their comfort zones. Olympic champions are always pushing themselves to new limits. For example athletes may attach weights to their bodies to put themselves under additional stress compared with what they will experience in competition in an attempt to increase their performance. Similarly scientists should experience working under difficult conditions to increase their performance. Working with less than asked for will push scientists to think, rationalise their experiments, search for alternatives, and innovate.

8 Researchers working in industry should get their boots dirty

I have no confidence in industrial research that has been conducted by researchers sitting at their laboratory benches or behind their desks. I cannot change something that I do not fully know about. When I was doing my research, I was most successful when I was working with the technicians, talking to them during their night shifts, being able to do their jobs if required. Because I understood their work, I was able to propose solutions to their problems. The same applies to clients. I was always more interested in visiting clients than attending a conference. By visiting clients and seeing how our company's product was applied, I was able to gain valuable insights for my company. If you manage a research team and your staff is doing no field work it is very likely your research will be useless.

But take care: researchers are not emergency staff. It is a common practice in industry to request researchers to work on short-term problems. In my experience these short-term problems soon become chronic problems in the production process. This is disruptive to researchers' routine. Usually these short-term contracts are requested by strong managers (i.e. those in the production area), who see little value in research. If you manage a research group you should not allow this to happen.

9 Researchers should be challenged

Good researchers will respond in a very positive way when challenged. If you are an executive, assess their behaviour when they are presented with a new challenge. If a researcher promptly starts excusing themselves, or contesting the challenge with poor arguments, move them out of your research centre. Researchers should respond positively to a challenge. If they decline a reasonable challenge, they are not right for industry. Good researchers must have the potential to change the industry.

10 Interdisciplinary research

If you work for the research centre of a mining company, invite a botanist to help you search for ore deposits. If you have a team of geologists, you may try to follow insect behaviour associated with volcano eruptions. Or if you work in a pharmaceutical company, you may want to invite a physicist working in the field of quantum mechanics to assist you with the development of an innovative medicine. I could provide endless examples of successful interdisciplinary research. The point is that innovation comes from thinking differently, making new associations, seeing the problem with a different perspective. Working across disciplines is a key to innovation in industrial research. To innovate in the way industrial research is conducted, research centres should collaborate with departments of universities working in different disciplines. Researchers should read more interdisciplinary journals (such as *Science*, *Nature*, *PNAS*) to capture new ideas. Papers published in interdisciplinary journals should be discussed with colleagues in your department. There should be a culture of inviting researchers from different areas to give seminars and explain the trends in their disciplines.

Being a successful researcher is not easy. You need to study and seek for the up-to-date, deal with a wide range of different problems in your entire professional life, and be responsible (which means at least being ethical and overcoming the roadblock of procrastination).

In my understanding the most important problem in science today is for scientists to solve their ethical problems and improve the way they practice. Every day, researchers should do everything possible to carry out their scientific work responsibly.

Good luck with your own research!

ENDNOTES

Chapter 1

1 'Scientists' and 'researchers' refer to the same professionals. I have not made a distinction between them.

2 Behaviourism, in psychology, justifies the use of the scientific method by means of observing human interaction with the environment. Psychology studies human behaviour through observation and measurement.

3 Sustainability is a form of development that meets the needs of the present without compromising future generations. This definition applies not only to industries but to any aspect of human activity. In industry it is understood as the organisation's commitment to develop its activities aiming at profit, respect to society and the environment.

Chapter 2

1 Webb LB (2006) The impact of projected greenhouse gas-induced climate change on the Australian wine industry. PhD thesis, University of Melbourne, Melbourne. For a comprehensive report on the impact of climate change in Australia read: 'Climate change in Australia: technical report' (2007) CSIRO.

2 Mears P (1994) *Quality Improvement Tools and Techniques*. McGraw-Hill, New York.

3 The impact factor provides a quantitative method for ranking, evaluating, categorising and comparing journals. The impact factor is a measure of the frequency with which the 'average article' in a journal has been cited in a particular year or period. The annual impact factor is a ratio between citations and recent citeable items published. It is calculated by dividing the number of current year

citations into the source items published in that journal during the previous two years.

4 The Digital Object Identifier (DOI) system is for identifying content objects in the digital environment. Information about a digital object may change over time, including where to find it, but its DOI name will not change. The DOI system provides a framework for persistent identification, managing intellectual content, managing metadata, linking customers with content suppliers, facilitating electronic commerce, and enabling automated management of media. DOI names can be used for any form of management of any data, whether commercial or non-commercial (e.g. scientific data). Using DOI names as identifiers makes managing intellectual property in a networked environment much easier and more convenient, and allows the construction of automated services and transactions.

5 Hirsch JE (2005) An index to quantify an individual's scientific research output. *PNAS* **102** (46), 16569–16572. doi: 10.1073/pnas.0507655102.

6 Web of Science is an internet-based service that provides researchers, administrators, faculty and students with quick, powerful access to citation databases. It covers over 10 000 of the highest impact journals worldwide and over 110 000 conference proceedings.

7 Chambers DS and Wheeler DJ (1992) *Understanding Statistical Process Control.* 2nd edn. SPC Press, Knoxville, TN.

8 Hill T and Lewicki P (2006) *Statistics: Methods and Applications: A Comprehensive Reference for Science, Industry, and Data Mining.* StatSoft Ltd, Tulsa, OK.

9 Couzin J and Unger K (2006) Scientific misconduct: cleaning up the paper trail. *Science* **312**, 38–43, doi: 10.1126/science.312.5770.38; and Service RF (2002) Bell Labs fires star physicist found guilty of forging data. *Science* **298** (5591), 30–31, doi: 10.1126/science.298.5591.30.

10 For details I recommend you access the websites of major publishers.

Chapter 3

1 1 tCO_2e is one ton of carbon dioxide equivalent. To read more about it I recommend the following report from The Intergovernmental

Panel for Climate Change: *Climate Change 2007, Fourth Assessment Report*, Working Group reports I to III (IPCC, Cambridge University Press, Cambridge, 2007); www.ipcc.ch/.

Chapter 4

1 The Mars Exploration Rover mission has the objective to determine the hydrological, climatic and geologic history of two sites on Mars where conditions may have been favourable to the preservation of evidence of pre-biotic or biotic processes. Kerr RA (2004) On Mars, a second chance for life. *Science* **306** (5704), 2010–2012, doi: 10.1126/science.306.5704.2010.

2 de Souza Jr PA and Brasil GH (2009) Assessing uncertainties in a simple and cheap experiment. *European Journal of Physics* **30**, 615–622. doi: 10.1088/0143-0807/30/3/018.

Chapter 6

1 There are many books available to help you improve your presentation skills. A good one is: Bradbury A (2006) *Successful Presentation Skills*. 3rd edn. Kogan Page, London, UK.

Chapter 7

1 I recommend the following papers (and references therein) for further information on transgression in science:

- Errami M and Garner H (2008) A tale of two citations. *Nature* **451**, 398–399. doi: 10.1038/451397a.
- Long TC, Errami M, George AC, Sun Z and Garner HR (2009) Responding to possible plagiarism. *Science* **232**, 1293–1294. doi: 10.1126/science.1167408.
- Couzin-Frankel J and Grom J (2009) Plagiarism sleuths. *Science* **234**, 1004–1007. doi: 10.1126/science.324_1004.

2 Couzin J (2008) Gene tests for psychiatric risk polarize researchers. *Science* **319**, 274–277. doi: 10.1126/science.319.5861.274.

Chapter 9

1 Lee M (2005) *EU Environmental Law: Challenges, Change and Decision-Making*. Hart Publishing, Oxford.

2 *IPCC Special Report on Carbon Dioxide Capture and Storage* (2005). Prepared by Working Group III of the Intergovernmental Panel on Climate Change (Eds B Metz, O Davidson, HC de Coninck, M Loos and LA Meyer). Cambridge University Press, Cambridge, UK and New York, NY.

FURTHER READING

Careers in science

Acheron C and Kickuth A (2005) *Make Your Mark in Science.* Wiley-Interscience, Hoboken, New Jersey.

Denholm C and Evans T (Eds) (2007) *Supervising Doctorates Downunder: Keys to Effective Supervision in Australia and New Zealand.* ACER Press, Melbourne.

Doherty PC (2005) *The Beginner's Guide to Winning the Nobel Prize: Advice for Young Scientists.* Melbourne University Press, Melbourne.

National Academy of Sciences (1995) *On Being a Scientist: Responsible Conduct in Research.* 2nd edn. National Academy Press, Washington, DC.

Philosophy and history of science

Beveridge WIB (2004) *The Art of Scientific Investigation.* Blackburn Press, Caldwell, NJ.

Bright Wilson E Jr (1991) *An Introduction to Scientific Research.* Dover Publications, New York, NY.

Descartes R (1994) *Discourse on Method.* Orion Publishing, London, UK.

Doherty PC (2008) *A Light History of Hot Air.* Melbourne University Press, Melbourne.

Gauch HG Jr (2002) *Scientific Method in Practice.* 1st edn. Cambridge University Press, Cambridge, UK.

Gribbin J (2003) *Science: A History.* Penguin Books, London, UK.

Medawar PB (1981) *Advice to a Young Scientist.* Basic Books, New York, NY.

Popper KR (2002) *The Logic of Scientific Discovery.* Routledge, London, UK.

Popper KR (1972) *Conjectures and Refutations: The Growth of Scientific Knowledge*. 3rd edn. Butler & Tanner Limited, London, UK.

Ramon y Cajal S (2004) *Advice for a Young Investigator.* 1st edn. The MIT Press, Cambridge, MA.

Sagan C (1997) *The Demon-Haunted World: Science as a Candle in the Dark*. Random House Publishers, Toronto, Canada.

Wolpert L (1993) *The Unnatural Nature of Science*. Faber and Faber Ltd, London, UK.

Young JW (1975) *A Technique for Producing Ideas*. NTC Business Books, Chicago, Ill.

Data management and analysis

Batini C and Scannapieco M (2006) *Data Quality: Concepts, Methodologies and Techniques*. Springer, Heidelberg, Germany.

Box GEP, Hunter WG, Hunter JS (1978) *Statistics for Experimenters.* John Wiley & Sons, New York, NY.

Grabe M (2005) *Measurement Uncertainties in Science and Technology.* Springer, Heidelberg, Germany.

Taylor JR (1997) *An Introduction to Error Analysis*. 2nd edn. University Science Books, Sausalito, CA.

Responsible conduct of science

Ahearne JF (1999) *The Responsible Researcher: Paths and Pitfalls*. Sigma Xi, The Scientific Research Society, Triangle Park, NC.

Mecrina FL (2005) *Scientific Integrity*. 3rd edn. ASM Press, Washington, DC.

Communication in science

Cribb J and Hartomo TS (2002) *Sharing Knowledge: A Guide to Effective Science Communication*. CSIRO Publishing, Melbourne.

Innovation

Byrd J (2002) *The Innovation Equation: Building Creativity & Risk Taking in Your Organization*. Jossey-Bass/Pfeiffer Publishing, San Francisco, CA.

Chakravorti B (2003) *The Slow Pace of Fast Change: Bringing Innovations to Market in a Connected World.* Harvard Business School Press, Boston, MA.

Davila T, Epstein MJ and Shelton R (2006) *Making Innovation Work: How to Manage It, Measure It, and Profit from It.* Wharton School Publishing, Upper Saddle River, NJ.

Dosi G (1982) Technological paradigms and technological trajectories. *Research Policy* **11** (3), 147–162.

Evangelista R (2000) Sectoral patterns of technological change in services, economics of innovation. *Economics of Innovation and New Technology* **9**, 183–221. doi: 10.1080/10438590000000008.

Fagerberg J, Mowery DC and Nelson RR (eds) (2004) *The Oxford Handbook of Innovations.* Oxford University Press, Oxford, UK.

Harvard Business Review on Innovation (2001) Harvard Business School Press, Boston, MA.

Mansfield E (1985) How rapidly does new industrial technology leak out? *Journal of Industrial Economics* **34** (2), 217–223. doi: 10.2307/2098683.

McKeown M (2008) *The Truth About Innovation.* Pearson/Financial Times, New York, NY.

Nelson RR and Winter SG (1977) In search of a useful theory of innovation. *Research Policy* **6** (1), 36–76. doi: 10.1016/0048-7333(77)90029-4.

OECD (1995) 'The measurement of scientific and technological activities. Proposed guidelines for collecting and interpreting technological innovation data. Oslo manual'. 2nd edn. DSTI, OECD/European Commission Eurostat, Paris. Accessed 30 June 2008 from: http://www.oecd.org/dataoecd/35/61/2367580.pdf

Scotchmer S (2004) *Innovation and Incentives.* MIT Press, Cambridge, MA.

Utterback JM and Suarez FF (1993) Innovation, competition, and industry structure. *Research Policy* **22** (1), 121. doi: 10.1016/0048-7333(93)90030-L.

von Hippel E (1988) *The Sources of Innovation.* Oxford University Press, Oxford, UK.

Wolpert J (2002) Breaking out of the innovation box. *Harvard Business Review* **August**.

Woodman RW, Sawyer JE and Griffin RW (1993) Toward a theory of organizational creativity. *Academy of Management Review* **18** (2), 293–321. doi: 10.2307/258761.

INDEX